QUANTUM PHYSICS

QUANTUM PHYSICS

FROM SCHRÖDINGER'S CAT TO ANTIMATTER

STEN ODENWALD

This edition published in 2022 by Arcturus Publishing Limited
26/27 Bickels Yard, 151–153 Bermondsey Street,
London SE1 3HA

AD008454UK

Printed in the UK

Contents

INTRODUCTION

'We must be clear that when it comes to atoms, language can be used only as in poetry.'

Niels Bohr

The 20th century will long be remembered as the century when we finally began to understand the basic structure of our physical world with its many laws and phenomena, and fashioned robust explanations for how things came to be the way they are. From the outermost limits of the observable universe to the innermost structure of the atom, the physical world resembles a bewildering set of nested Matryoshka dolls. The largest ones encompass the entire cosmos of distant galaxies while the smallest ones encapsulate the scale of the atom and beyond. Between these lie the medium-sized realms, which span the scale of our familiar day-to-day world.

The nature of the human-sized and cosmic-scale worlds seems entirely straightforward and is circumscribed by how ordinary matter operates under a small number of forces, especially gravity. The ways in which matter moves through space and time has been largely codified through three centuries of experimentation and theoretical investigation and termed 'classical physics'. At the same time, a collection of phenomena was discovered that could not be explained by classical physics alone. Something was missing that prevented an understanding of why heated iron glows red, and atoms emit specific wavelengths of light and no others. Only by following these clues and investigating the physical world at atomic and sub-atomic scales could, at last, these phenomena be accounted for; however, there was a price to be paid. Entirely new laws and theories unlike those in classical physics had to be defined and crafted in what became known as 'quantum mechanics'.

Some ideas from classical physics could be carried over into the quantum realm but they invariably had to be expressed in a new language or even re-interpreted entirely. For example, the conservation of energy was still a good law of nature, but in quantum mechanics it had to be qualified by an uncertainty principle that said it could also be violated depending on how long you measured a system. In classical physics, a wave was

A segment of the 27-km beam line of the Large Hadron Collider at CERN.

a wave and a planet was a planet, but in the quantum realm, objects could display both wave and particle attributes though not simultaneously. In classical physics, when you observe a rock or a dust grain, it remains the same before and after you observe it. But in quantum mechanics, the very act of observing an object changes its state. In fact, there is no actual reality for an object at all. Its wave or particle aspects come into existence at the moment of measurement when your experiment asks a specific question such as *What is your mass?* (particle) or *Which of the two slits did you pass through?* (wave). There were also some ideas in classical physics that could not be explained within this framework alone, although it was still possible to make accurate predictions even with this lack of knowledge.

Whether you are calculating the electrostatic repulsion between two like-charged particles, or the movement of a planet under the influence of the sun's gravity, classical physics gave you the detailed mathematics to do so, but not an explanation for the origin of this odd action-at-a-distance effect. How can two bodies exert an influence on each other through empty space without touching? Within a few decades of the advent of quantum mechanics, the nature of how forces operate was at last uncovered and this, too, was revealed to be at its core a quantum phenomenon known as a quantum force field or simply a quantum field. But to make this theory of quantum fields work, the very nature of empty space had to be entirely re-interpreted and made mathematically rigorous. This work was begun with the reformulation of quantum mechanics so that it was consistent with Einstein's relativity. A new ingredient, antimatter, appeared in the mathematics along with a vacuum 'state' populated by virtual particles with insufficient energy to be real particles. It was these ghost-like particles that, when exchanged in large enough numbers, produced the forces we experience.

As new technologies for probing the composition of matter were developed by the mid-20th century, a zoo of new forms of elementary particles appeared, necessitating the creation of new

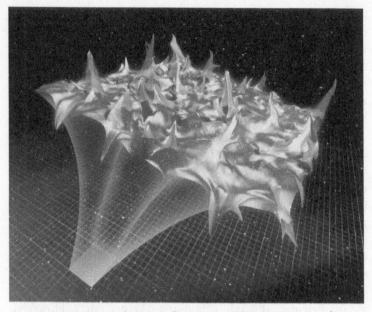

An artist's impression of space at the quantum 'Planck Scale'.

theories for nuclear forces and their constituent particles. The quark-gluon model and the electroweak model defined the new Standard Model for atomic physics phenomenology. Quantum mechanics, meanwhile, evolved with new mathematical techniques to explore the deep structure of space, time and matter – pursuits well beyond any that the developers of quantum theory had imagined in the 1920s. Today, the Standard Model appears experimentally perfect, although chinks in the mathematical formulation of the Standard Model seem more like gaping holes as time goes on. The biggest challenge is to find a quantum field theory for gravity, but gravity is linked to the structure of space and time. This pursuit of a quantum theory of gravity is nothing less than a theory for the origin and nature of space and time themselves.

PHYSICISTS ON QUANTUM MECHANICS

Niels Bohr famously noted 'If you can fathom quantum mechanics without getting dizzy, you don't get it.' *On the other side of The Pond, physicist Richard Feynman remarked,* 'Do not keep saying to yourself, if you can possibly avoid it, "But how can it be like that?" because you will get "down the drain", into a blind alley from which nobody has yet escaped. Nobody knows how it can be like that.' *These are not the comments by physicists who are simply trying to brush off questions by students or the general public. These are the comments by the exasperated architects of quantum mechanics that reflect the deep mystery of why our world is cleaved in this way, with mind-numbing consequences we will explore in the chapters to follow.*

CHAPTER 1:

The World as We See It

Our familiar world is not quite as it seems. Mountains, waterfalls and the ground beneath our feet seem solid enough, but are in fact built up from even simpler elements that we cannot see without the aid of technology. Most of human history has been involved with the exploration and utilization of these different forms of matter, beginning with stone tools and the making of bronze. These rudimentary ideas evolved steadily over the millennia into a pragmatic understanding of chemistry in the hands of medieval alchemists. Meanwhile, rudimentary classification schemes that served the Ancients well enough gave way to more complex ones as the number of elementary forms exploded in numbers. Copper, silver, gold and salt begged for a refinement of the Aristotelian category of earth into a finer-grained concept of terrestrial matter. This led to various attempts to systematize a growing number of substances in terms of basic common properties such as density, chemical reactivity and colour among other features.

Our common experience of the world

Go outside and take a look around: what do you see? Apart from the trappings of modern technology with its buildings, power lines and mobile phone towers, the background world of rocks, soil, water and air have been seen by humans since the dawn of the species. In fact, these same elements were the common ingredients of the natural scenery perceived in one way or another by every living creature on earth for much of the last 500 million years. For most of this time, organisms regarded these elements of their world with a disinterested eye for those that had them, but navigated the world as best they could for survival. Even for our most primitive ancestors up until the dawn of our current species, *Homo Sapiens Sapiens* some 40,000 years ago, there was little recorded curiosity about why rocks were different from water, and what the nature of air and fire were. These ingredients of the world were simply what they were, with some noteworthy variations. There was drinkable water, brackish water and salty water. There were rocks of every different colour and hardness. And sometimes the invisible air was in motion on a windy day, or took on an awful stench from decayed flesh, or a beautiful fragrance from the flowers in the field. So our ancient ancestors knew of many different varieties of rock, water and air, and no doubt had names for the ones that mattered to them and their immediate survival. There was no particular reason to be more curious about the contents of the physical world other than to list those ingredients by name that had some utilitarian purpose. Among the 'rocks', for example, the most important became flint, gold, copper, tin, iron and salt. There were a variety of plants that were ultimately named such as wheat and barley, as well as animals like sheep, horses and goats as well as predators. There were different names for lightning storms than for tornadoes, and perhaps even for the different kinds of clouds in the sky as well.

In addition to the rocks, air and water in our world our remote ancestors may have noticed two other ingredients that behaved very differently. One of these was fire, which seemed to be produced when friction heated up a solid substance to high enough temperatures, especially organic materials like wood. Fire is similar to air in that it has no obvious density, and it flows upward from the ground into the sky just like air does. The sun has much in common with fire in that it produces light and heat unlike all the other objects in the sky. In fact, objects in the sky are definitely viewable on a nightly basis but stars, planets and the moon do not resemble either rocks, air, water or fire so they were considered to be made from some new Fifth Ingredient. So with these five ingredients we can pretty much classify all the things we see in the commonplace natural world.

The hidden world just beyond our senses

The largest things we can see in the world are most definitely dramatic. Spectacular mountain vistas, dramatic storm events from lightning to tornadoes, and waterfalls and ocean breakers abound only a few days' walk from just about anywhere you are. But seemingly as a fabric to this grandeur one can encounter a new landscape of the very small that is nearly as rich in novelty. A walk on the beach, or at the base of a rocky cliff, brings you into contact with sandy particles that can in some instances be so minute the eye strains to focus on them. No two sand grains are precisely the same in shape or colour. The bewildering variety of rocky forms we see among the mountains now has an equal resonance at the scale of sand grains.

In some quarters of the world, these sand grains pile one atop the other to form sand dunes that resemble the more-distant mountains that barricade their movement. The visual confabulation between the solid rock of the mountains and the delicate shapes of the sand dunes calls out for us to imagine that the underlying structure of mountains is nothing less than a more solid form of the sandy dunes. We are not at a later time disappointed to discover that solid rocks can in fact be ground by friction into grains of sand-like particles. We also discover that the variation in the sand grain colour is a reflection (quite literally!) of the differing minerals we can collect on the mountainside. But rocks are not alone as ingredients to be pulverized into another granular form. Air laden with invisible water vapour can transform into visible water droplets and hail, while water itself can dissolve into a spray of particles down to the limits of human vision. It seems that all of the macroscopic forms we observe so easily in nature, rocks, air and water, can be dissolved into finer and finer particles. But why stop here? What is it that limits human vision and prevents us from experiencing an even smaller world below the size of sand grains?

FLOATERS

Another way to experience the world of the very small without the help of technology is through the visual phenomenon called 'floaters'. Red blood cells as a result of haemorrhage, and white blood cells as a result of inflammation, are common types of cellular material trapped inside the transparent vitreous humour of our eyes. Since the largest of these cells are about 20 micrometres (0.02 mm) in diameter, they shadow the light arriving at the retina, and you experience these shadows as large, annoying, floating spots in your visual field.

It is not a matter of happenstance whether you can see small things or not. The human eye is a lens whose magnification is organically limited. For viewing distant, large scenery, its shape is a very thin double-convex lens. But for close-in viewing, the ciliary muscles in the eye pull at the edges of this lens to make it fatter in the middle. This allows you to focus at will on nearby objects. The limits to human vision set by this lens are such that objects 0.04 mm wide (the width of a human hair) are just distinguishable by good eyes, but objects 0.02 mm wide are not. The dot at the end of this sentence is about at the limit of what the human eye can discern. You will notice that as you keep staring at the full stop, its shape seems to change almost imperceptibly. This happens because the rod and cone cells in the retina have finite sizes like the squares on a chessboard. The image of the object is so small that when the lens focuses it on the retina the image straddles only a few rods and cones. The shape of the full stop along its edge is no longer smooth, but varies in a jagged manner as your eye motion carries its image across the chessboard of our retinal rods and cones. For poppy seeds on a bagel, their shapes seem well-formed at just under 1 mm because the image of a poppy seed covers several dozen rods and cones on the retina giving you the perception of a stable, round shape. Similarly, if you scatter salt grains on a piece of black paper, you can also just make out their shapes at a scale of about 0.3 mm. But, if you perform the same experiment with very fine beach sand at a scale of 0.02 mm, you will visually sense that the grains exist but their shapes will appear indistinct and variable when you try to stare directly at one individual grain.

STUDYING THE SMALL

Early investigators of nature understood there was a finer-grained world to perceive but realized that much of its details were hidden well below the limit to human visual acuity. One of the earliest tools developed perhaps accidentally to get around this problem was the Nimrud Lens of 750 BC developed by the Assyrians. It was a rudimentary glass lens that could magnify about three times. Whether it was actually used to study small details is unknown because it also resembled decorative glass used for other purposes. The ancient Greeks and Romans are generally credited with having filled glass spheres with water to serve as rudimentary magnifiers. During the 1st century AD, *Seneca the Younger noted that* 'Letters, however small and indistinct, are seen enlarged and more clearly through a globe or glass filled with water.' *There were many discussions during the Middle Ages about how light could be reflected and refracted, but it wasn't until the 11th century that so-called Reading Stones were invented and systematically used by monks to illuminate their manuscripts. These were glass spheres cut in half to form a lens with modest magnification. Then, by 1286, the first eyeglasses were invented using glass lenses polished to the correct shapes.*

The world through the microscope

As inventors continued to experiment with lenses of modest magnifications from 3 to 10x, eventually the first microscopes were developed. As you make the curvature of the lens more extreme, the lens is able to increase its magnification. A big glass sphere has a rather gentle surface curvature, but very small spheres allow for very extreme curvatures. The Dutch scientist Antonie Van Leeuwenhoek (1632–1723) used this curvature principle to fabricate tiny spheres of finely polished glass mounted in a frame to create simple viewers with magnifications of 250x and more. The invention of high-magnification lens systems dramatically opened up a new world for what would become the new science of microscopy. Virtually anything from fleas to sand were scrutinized and artistically drawn. At these magnifications, red blood cells could be studied, as well as the vast armada of microscopic 'little animals' discovered in a drop of pond water by Leeuwenhoek himself. Before the advent of photography, microscopists had to be expert artists able to draw the finest details in the correct proportions as they hunched over a microscope by candlelight for hours at a time.

Walther Flemming's drawings in the 1880s from his systematic study of cells eventually revealed the sequence of steps in cell division and reproduction. The drawings were an art form in and of themselves, revealing details as small as 0.01 mm. At magnifications of over 1,000x, the limit to optical microscope design, the world explodes into a level of fine structures inside cells themselves. Flemming's almost imperceptible chromatin fibres that were essential to cell division and shepherded chromosomes during cell mitosis, were only 10 microns (0.01 mm) long but 0.010 microns (0.00001 mm) in diameter. A single human hair (75 microns or μm), barely visible to the eye would be

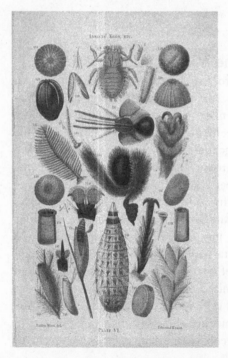

This drawing was printed in 1867 and shows microscopic details in a variety of objects.

7,500 times as large. Not only were biological systems of intense curiosity to 19th century microscopists, but a whole host of rock crystals, sand and dust grains suddenly resolved themselves into a bewildering variety of shapes and compositions.

ANTONIE VAN LEEUWENHOEK

Antonie van Leeuwenhoek (1632–1723) was born in Delft, in the Dutch Republic. His father, a basket maker, died when he was only five years old; his mother, who was wealthy, later re-married but his stepfather died when he was ten years old. He was sent off to live with his uncle, an attorney in Benthuizen and at age 16 became a bookkeeper's apprentice at a linen-draper's shop in Amsterdam.

This is hardly the pattern taken by many other contributors to science, but nevertheless Antonie arrived as the Father of Microscopy from a very practical consideration: he wanted to inspect the quality of the threads being used in his fabrics. So he learned how to make microscopes following instructions gleaned from previous inventions in optics at the time. Instead of grinding lenses he discovered that melting glass in a flame created spheres that produced very high magnification when sighted through them. This discovery allowed him not only to inspect fine threads, but opened up an entire world of microscopic nature only dimly viewed by his more scientific contemporaries.

Over the course of his investigations, he published no fewer than 190 letters to the Royal Society announcing his many discoveries of microscopic life. He was elected to the Royal Society in 1680, which was an honour that caught him completely by surprise. Van Leeuwenhoek was so advanced in his field of microscopy that famous contemporaries such as Robert Hooke are said to have opined that the field had come to rest almost entirely on one man's shoulders.

During his life a parade of famous world figures would in turn squint through his devices, including Gottfried Leibniz, William III of Orange and his wife Mary II, as well as Tsar Peter the Great.

Antonie van Leeuwenhoek.

Optical microscopes, meanwhile, began to show their limits after several centuries of use and refinement. Their ability to resolve details finer than 0.01 µm was hampered by the very wavelengths of light they were using to form their images even with the best of lenses. The newer technologies of the 20th century led to breakthroughs in microscopy with the invention of electron microscopes (1931) and scanning tunnelling microscopes (1980s). For the first time, the structure of the smallest life forms, the viruses, became visible at magnifications of 50,000x, impossible to achieve with light, but routine achievements using electrons as 'bullets' to form the images. The micron (0.001 mm) was replaced by the nanometre (1 nm = 0.001 µm) as the new unit of practical measure. At a magnification of 100,000x the terrifying polio virus that had caused so much misery among humans, was resolved for the first time in 1953 as enigmatic, round particles only 20 nm (0.00002 mm) in diameter.

Below a scale of 1 micron, bacteria and viruses began to show their detailed shapes. Among the most numerous life forms on earth, there seemed to be no limit to their varieties. Nature had not as yet run out of innovative ways to package living matter. But among objects in the inorganic world dominated by rock crystals, we encounter a much more impoverished world. There are only a maximum of 32 different ways that crystals can form in three-dimensional space, called the crystallographic point groups, which forces rocks to take on only a few novel microscopic forms.

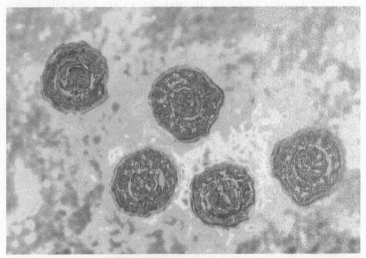

Hepatitis C virus viewed with a transmission electron microscope. The actual viral diameter is around 22 nm.

A glimpse of the atomic world

At 1 nanometres, which is one billionth of a metre (1 nm = 0.000001 mm), even electron microscopes began to reach their practical limits. Just in time, a new technology called field ion microscopy was invented in 1951 that opened the door to the ultimate scale of natural structure. Unlike a sleek conventional microscope, the field ion microscope looks more like a random collection of vacuum vessels, stainless steel pipes and electrical cables with no obvious imaging 'lens'. But deep within the tangle of this technology, currents of electricity and needles honed to unimaginably fine points combine to create breathtaking scenes a million times finer than human vision could allow. Features as small as 0.1 nm revealed a world populated by individual balls of matter fixed in geometric lattices. Science and technology had at last reached the limits to matter and we could watch as solid bodies literally ran out of substances as we know it. At one scale, a sheet of copper looked much as it does as you hold it in your hand. At a scale of 1 nm, it dissolves into a geometrically-precise collection of dots leaving no hint of its normal appearance at the human scale. What was also apparent was that the variety of forms we had encountered at every stage of magnification had started to wane. Among the features and forms apparent at 1 nm and smaller, a great reduction in the number of unique shapes had started to occur. The collections of atom-sized dots led to arrangements that seemed somehow less complex. Studies of both organic and inorganic samples revealed the dim shapes of the individual constituent molecules and atoms, but the overall forms were not nearly as varied as the bewildering numbers of

Historic photograph of tungsten needle atoms viewed by a field ion microscope.

shapes encountered by early microscopists like Leeuwenhoek as they studied organisms in pond water.

The technology for seeing small things did not cease evolving after the field ion microscope appeared in 1953. A few short decades later another technique was invented in 1981 called scanning tunnelling microscopy. Its inventors, Gerd Binnig and Heinrich Rohrer working at the IBM Laboratory in Zürich, received the Nobel Prize in Physics for their path-breaking work in 1986. Even clearer images than possible with the field ion microscope were now possible. The sample you were studying did not have to be ground into a fine needle, but instead a needle-like sensor could be scanned across the sample to build up an image of what was beneath the needle's atom-sized tip. Like a precocious child given a microscope for Christmas, engineers and scientists placed a whole host of objects under the scanner's needle. Most of these were stable mineral samples but eventually even human DNA took centre stage, revealing for the first time its double-helix shape.

The journey from the familiar macrocosm of mountains and waterfalls to the sub-microscopic world of atoms took millennia to complete, but we now see the entire world much more clearly. At its foundation are small objects called atoms that seemingly link together in various combinations to create all of the structure we see around us. These are not theoretical objects as they had been for over 2,000 years, but actually observable features of our world with the help of a little technology. Another thing that we notice is that movement seems to be suspended. With optical microscopes we can still see organisms and cells moving about but with some difficulty. This is a feature of a property of matter called viscosity. You experience some of its effects when you go for a swim. Viscous forces working against movement are annoying for 2-metre-tall humans, but become increasingly worse as the size of a system decreases. Most microscopic living systems seem suspended in space as viscous forces lock them into a dynamic prison making movement difficult. To them, ordinary water feels more like molasses. Some animals such as the paramecium, which is only 50–300 microns in size, have hair-like organisms on their surfaces called motile cilia. Like oars on a ship, cilia move the paramecium about in search of food. Over 40,000 species of organisms have evolved cilia as a means for moving about. Nevertheless, there is nothing about the motion or lack of motion that we see at these scales that suggests anything new in the way nature moves matter about under self-propulsion, buoyancy or gravity.

Despite having resolved the world into what seem to be its most elemental atomic structures, there is very little that can be gleaned about how atoms work from looking at even the highest resolution images provided by field ion microscopes and scanning tunnelling microscopes. For all the diversity of the larger structures in the world, these imaged atoms all look much the same, one to the other. We can discern the fuzzy shapes of some large molecules, but as for their constituent atoms, they have similar sizes (0.1 nm) and shapes (round). There is no obvious clue in even the best images that tell us how gold atoms differ from carbon atoms, or how they actually are linking together to form their collective shapes. For this, we have to approach the study of matter from a different angle entirely. That journey was begun in India, China and Greece over two millennia ago.

CHAPTER 2:

Atoms and the Void

When it comes to investigating the natural world, it is very tempting to re-write history to make the discoveries and insights flow more logically. You can imagine that, at first, our ancestors looked at big things like rocks and then studied smaller things such as sand to conclude that rocks are made from sand. Then when they developed lenses and microscopes of every size and variety, they discovered that rocks are made from mysterious, small balls of elementary substance. These nuggets of matter could now, at last, actually be seen using field ion microscopy in the 20th century, which I described in Chapter 1. Once these kernels of matter were spotted, the basic laws of chemistry could be laid out because you could now see how these balls of matter combined to create new compounds. This logical rewrite is, however, completely false. Humans are the quintessential multi-taskers of the animal kingdom, capable of going off in many directions at the same time. The realization that every big thing we could see with the naked eye was built upon elementary 'atoms' that we could not see was an idea dating from at least the first millennia BC when we didn't even know what chemistry was.

Introducing the atom

By the 5th century BC, it was ancient Greek philosophers such as Democritus and Leucippus that considered this 'atomic' possibility, but the sages and gurus on the sub-continent of India had already reached this realization at least three centuries earlier. The Vedic sage Kanada reflected on this possibility by mentioning 'particles too small to be seen mass together into the substances and objects of experience'. He called them *paramanu*, which was Sanskrit for the smallest possible division of matter. There were four of these, earth, air, fire and water, along with five elementary substances, ether, time, direction, mind and soul, which combined to create everything in our particular corner of reality. In China, however, the idea that there were 'uncuttable' atoms in the world was not at all a part of their philosophical world-view until centuries after the ancient Greeks.

What these ancient philosophers came up with was the idea that all forms of matter were the result of the assembly of vast collections of small spheres. Democritus called these *atomos*, meaning 'undivided', which is the term we use today. Each of these atoms was indivisible but had the properties of the final substance. For example, if you were holding a lump of pure gold in your hand, you could divide that lump into smaller and smaller pieces until you reached the size of the atoms of gold themselves. At this point, each atom would have identical properties of gold such as its colour, but could not be divided further. There was another feature of this idea that was also absolutely necessary. There had to be places where there were no atoms at all called the Void. Without the Void, atoms would be locked in a massive traffic jam unable to move about. So the Atomist School not only proposed that atoms existed but that there existed an absolutely empty Void.

What was unique about both the ancient Greek and Vedic philosophers was that observation and logical deduction were an important ingredient to how they came up with some of their ideas. They also worked by using analogies. The idea of atoms no doubt came from simple observations of the finer structure to the things around them such as grains of rice or sand. But as sensible and logical as it seemed, even atomism was not without its powerful detractors. The legendary philosopher Aristotle, for example, was very influential in promoting his Four Elements view of nature and not believing in the idea of the indivisible atom, which he considered a logically flawed idea. Aristotle's idea of Earth, Air, Fire and Water was later extended to include the Fifth Essence (the Quintessence) called Aether; the luminous and perfect substance of the stars and planets. Even ancient Chinese scientists used Aristotle's Four Elements scheme to organize their observations. Atomism, however, was still considered an intellectual aberration in the Western world well after the Middle Ages over 1,000 years after it was first proposed. But even more disturbing was the idea of the empty Void that Aristotle dismissed in his famous quotation '*horror vacui*', translated later as 'nature abhors a vacuum'. Even apart from atoms, there were many acrimonious debates over the existence of a Void. By the 13th century, the Roman Catholic Church was fully on board with Aristotle's masterful survey of the physical world and his Elements. In some places, such as France, to propose the existence of an empty Void was an excommunicable offence irrespective of the impeccable logic you might use to support the idea. But whether you believed in atoms and the Void or Aristotle's Five Elements, the practical consequences were much the same: there simply were none!

Atoms partake of simple chemical combinations
Atomic theory did not help your local jeweller or metallurgist fabricate and purify gold and silver. Atomic theory, or even

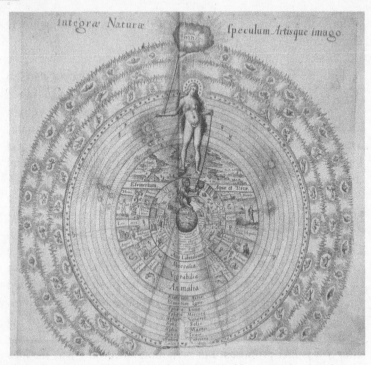

Aristotle's elements from a book by Robert Fludd c. 1624 showing the unity of nature.

Aristotle's elemental version of matter, did not help create bronze by informing you how to combine copper and tin in the right proportions. All of the practical chemistry going on between the time of ancient Greece and the 16th century Renaissance was conducted by trial and error. Notes were taken of procedures and combinations of compounds that worked towards a specific end. These precious recipes were usually written down in secret writings or memorized as incantations. Meanwhile, along the way, ancient Chinese chemists discovered gunpowder, and others in the West discovered the basic processes of distillation, condensation and purification. Ancient Greek ideas about atoms, along with Aristotle's Elements, were suppressed during the Dark

Ages between the 5th and 15th centuries, but were available in the Arabic-speaking world where *Al Kimiya* was being intensively studied. Over the course of centuries, alchemists, as they were called, succeeded in creating hundreds of interesting compounds, but their true goal was to transmute lead into gold, or to create elixirs and potions that would extend life. By the early 1700s even Sir Isaac Newton tried to synthesize 'philosophical mercury' using alchemical techniques and knowledge.

Basic alchemical symbols for compounds.

Alchemy seems like a pointless subject to us today, but along the way to its unachievable goal of turning lead into gold, the practitioners discovered many chemical compounds such as *vitriol* (sulphuric acid) and *aqua vitae* (ethyl alcohol). They also found that nature was not haphazard. There were specific steps in carefully prescribed sequences that were needed to create many compounds reliably. What they also discovered through centuries of trial and error was that some 'base' ingredients such as gold, copper, lead, sulphur and iron resisted being reduced into even more elementary ingredients. For instance, they knew that the compound cinnabar could be heated to produce puddles of liquid mercury and sulphur, but no amount of further effort converted mercury into still more basic ingredients. While alchemists viewed this list of stable base elements as something of a curiosity, eventually this list became the starting point for others to investigate whether there were more of these elements in nature and exactly how they entered into chemical reactions. By 1648, it was the Anglo-Irish polymath Robert Boyle who discovered the chemical conservation of mass, and published an important book *The Sceptical Chymist* in which he discussed the distinction between chemistry and alchemy. From now on, these pursuits would go their separate ways with alchemy still rooted in millennia-old mystical ways to create gold and fabled elixirs. Chemistry, on the other hand, would employ the new 'scientific method' championed by Sir Francis Bacon to systematically and carefully investigate how elements combined to create compounds. Along the way, entirely new elements such as hydrogen, oxygen and cobalt were discovered.

By the turn of the 19th century, the French chemist Joseph Proust discovered that elements always combined in simple whole-number ratios to form new compounds such as water being two parts of hydrogen to one part of oxygen. The English chemist John Dalton developed the first new model for atoms since the time of Democritus and explained why elements are chemically stable. His list of elements now included 20 known examples at the time, but would quickly grow as Dalton's atomic theory caught hold and investigators searched tirelessly for more examples in nature of these irreducible elements. By the end of the 19th century, the list had grown to over 83 examples.

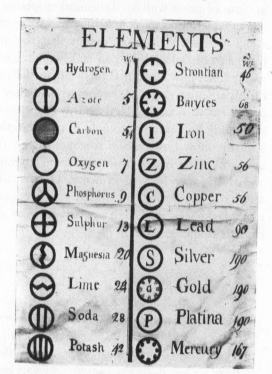

ELEMENTS

	w.		w.
Hydrogen	1	Strontian	46
Azote	5	Barytes	68
Carbon	54	Iron	50
Oxygen	7	Zinc	56
Phosphorus	9	Copper	56
Sulphur	13	Lead	90
Magnesia	20	Silver	190
Lime	24	Gold	190
Soda	28	Platina	190
Potash	42	Mercury	167

Dalton's list of the basic elements.

Clues to atomic structure

The work by Dalton to investigate chemical processes, along with his atomic theory, led him to the practical idea that atoms were linked together with what he considered 'hooks and eyes' to form compounds. There were only a few of these hooks and eyes on each atom and, for Dalton, this explained why they could only combine in small numbers with fixed integer ratios. This idea became very popular among chemists of the 19th century and led to the ball-and-stick models for atoms and compounds that are used today in 21st century chemistry classes around the world.

Dalton had opened the door to thinking about the elements in terms of atoms with fixed chemical properties, but this atomic model was still no more provable than some of the older ideas about Aristotle's elements, except that the atomic theory in the hands of Dalton led to an understanding of how chemistry worked. But how was it that these minute balls of matter carried their particular chemical attributes? Did they really have the equivalent of hooks and eyes on their surfaces as Dalton had proposed? Because atoms could not be directly observed even with the most powerful microscopes of the 19th century, the nature of atoms had to be inferred from laboratory experiments and from other lines of evidence. Once again, scientists can multitask, and through an entirely separate line of investigation new phenomena were being uncovered that provided missing clues to atomic structure.

POINTILLISM

The practical idea used by chemists, that nature was composed of atoms, did not just remain within the laboratory as an organizing idea, but infiltrated the artistic mood of these times, resulting in an entire school called Pointillism. If nature was made from atoms, why not represent natural landscapes and subjects as collections of points on the canvas? In this instance, the 'atoms' were points of the three primary colours, which when viewed from a distance would blend to form the intended combined colour. It was a novel technique that led to some remarkable and famous compositions by Georges Seurat and Paul Signac in the late 1800s.

One curious property of matter was that, when you rubbed certain materials with a cloth, you generated what became known as static electricity. This phenomenon had also been known with amusement to the ancient Greeks, and specifically by Thales in *c*. 600 BC. The 18th century was a dramatic time for studying electricity and inventing batteries called Leyden jars to capture it. Lightning was discovered by the American polymath Benjamin Franklin as a form of electricity in motion. By the 19th century, the basic laws of electrical currents and magnetic fields were uncovered. These discoveries almost instantaneously led to the invention of the electric telegraph, and the electrical turbine that later went on to transform a coal-based world of steam engines into the electrified world we know today. But what was electricity? Initially some scientists thought it was a fluid generated by matter, but by the mid-1800s this fluid idea was abandoned, thanks to experiments with devices called Crookes tubes. Electricity was now a flow of moving electric particles that carried individual charges. These particles were later called *electrons* by Irish physicist George Stoney because each of these 'electric ions' seemed to carry an irreducible elementary unit of charge no matter how he set up his experiments.

A Crookes tube and electron beam.

The relationship between electrons and atomic structure was not suspected for many decades even as electricity was being intensively investigated. Eventually, a crude idea emerged that atoms were composed of electrons and some other positively charged particles to make the atom electrically neutral. This simple model was finally proposed by the English physicist J.J. Thomson *c.* 1900 and called the Plum Pudding model. The basic idea was that each atom was a sphere of positively charged particles in which the negatively charged electrons were embedded. Using the Crookes Tube or other devices, these electrons could actually be liberated from the atoms to create 'cathode rays'. It would be another 20 years, however, before anyone realized that atomic electrons play a key role in chemistry and make possible the bonds that tie elements together into compounds. Dalton's hook-and-eye idea would finally be translated into the new language of electrons.

Electrons and atomic nuclei

Within a few decades during the turn of the 20th century, the model for atoms quickly evolved. Ernest Rutherford at the University of Manchester in England devised a clever experiment to probe the internal structure of atoms without needing to see atoms at all! By 1911, new elements were discovered that were 'radioactive'. They actually emitted particles such as electrons (beta rays) and helium atoms (alpha particles). Rutherford's experiment would be to bombard a piece of thin gold foil with these alpha particles and see what happens. If atoms were just amorphous balls of well-mixed electrical dough, the alpha particles would shoot right through them and not be deflected. But what he found was that sometimes these helium atoms would be ejected back towards the source, which could only happen if atoms had some kind of denser, solid kernel of matter at their cores. Instead of a Plum Pudding model, Rutherford's experiments showed what he interpreted as a core of positive particles surrounded by a cloud of negative electrons.

There was, however, one important fact not included in this 1911 model that would quickly call it into question within a few years. Back in the 1800s, a very complete, mathematical theory of how electrons and currents operated was developed by James Clerk Maxwell. What he discovered in his equations was that a charged particle that is accelerated by some force will emit what he called 'electromagnetic' waves. These waves would carry off energy. Inside Rutherford's new atoms, the positive nuclear charge would create a force on the negative electrons that would cause them to accelerate and lose energy. In literally the blink of an eye, Rutherford's atoms should all collapse and the infalling electrons instantly neutralize the positive nuclear charges. If Dalton's atoms were created along the lines of Rutherford's model, there would be no atoms in nature. Either Rutherford's experiments were flawed, leading to the wrong model for the atom, or Maxwell was wrong and something was causing charged electrons inside an atom not to emit radiation and lose energy.

ERNEST RUTHERFORD

Ernest Rutherford (1871–1937) was born in Brightwater, New Zealand, and attended Canterbury College. In 1895 he won a research fellowship to the Cavendish Laboratory in England to continue his studies for a PhD in physics under the guidance of J.J Thomson. In 1907 he became a Professor of Physics at the Victoria University of Manchester, and was knighted in 1914 and replaced Thomson as the Director of the Cavendish Laboratory.

His early work in physics involved long-wave radio communication, but since 1897 he had worked on issues related to radioactivity. He coined the terms alpha- and beta-rays in 1899 to describe the two kinds of penetrating radiation given off by certain atoms. He also invented the concept of half-life to describe radioactive decay. The 1908

Nobel Prize in Chemistry was awarded to him 'for his investigations into the disintegration of the elements, and the chemistry of radioactive substances'. Along with Hans Geiger and Ernest Marsden, in 1909 he used alpha rays to probe the structure of gold atoms and concluded from this path-finding work that atoms have a dense nucleus and are not a uniform 'plum pudding' of positive and negative charges. Rutherford later went on to propose the neutron as a new constituent of atomic nuclei shortly before his death in 1937.

Ernest Rutherford.

Around this same time there were other models for atoms being considered. For instance, in 1902 the Cubic Model was proposed by Gilbert Lewis. This model was actually quite clever because electrons were now located at the corners of atomic cubes. It was able to explain single, double and quadruple chemical bonds but not triple bonds, so it was abandoned.

Even before the 1900s there had been many other proposals for atomic structure including Lord Kelvin's Vortex Atom of 1867, which was very popular in England through much of the late 1800s. Atoms were imagined to be spinning vortices in the luminiferous ether, whose motions and properties were dictated by fluid mechanics. There were plenty of examples of vortex motion in nature and these provided a potent visual analogy for atoms.

Although the vortices of Lord Kelvin were too floppy to account for the precise regularities of Dalton's chemistry, this idea led to a knotted version of vortex structure that did have some built-in regularities that could be mined. For example, knots can be classified into specific geometric families that do have

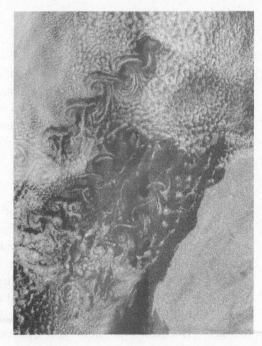

Von Kármán vortices in the clouds above the Canary Islands seen on 20 May 2015.

integer relationships. But vortex atoms are inherently unstable to dissipation and so the fundamental character of atoms, their immutability, could never be explained this way.

Even earlier ideas about matter had been considered, such as the model by the astronomer Roger Boscovich in the late 1700s. His basic idea, shared by philosopher Immanuel Kant, was that matter is not an essential ingredient to the world but is created through the conflict between forces. Matter is what Kant referred to as an 'epiphenomenon'. Boscovich's famous treatise *Philosophiæ naturalis theoria redacta ad unicam legem virium in natura existentium* (*Theory of Natural Philosophy derived to the Single Law of Forces which exist in Nature*) published in 1763 described in detail how this idea was supposed to work, but the effects were spread out in space and not localized into finite atomic forms, so this idea was also abandoned.

SATURN THE ATOM
In 1904, the Japanese physicist Hantaro Nagaoka came up with an atomic model based quite literally on the planet Saturn with its rings. Even Rutherford liked this model of a massive planetary core with orbiting electrons, but the enormous electrostatic forces between the electrons would have exploded the rings, so this model was also abandoned.

Quantization of electron orbits

By the time we arrive at the year 1913, Rutherford's atom was the best idea in circulation but its one fatal flaw was that it could not explain the stability of atoms. By some unknown mechanism, electrons within an atom would have to be kept from emitting energy and collapsing onto the nucleus. There did not seem to be anything in Maxwell's theory that would prevent this, so the German physicist Niels Bohr took the radical step of proposing two new principles in physics. First, electrons only orbited the atomic nucleus in specific paths just as the planets in our solar system do. Second, while they are in these specific orbits, they simply do NOT emit any energy at all! The first principle wasn't a completely outrageous idea because our solar system contains planets orbiting a massive sun, and the force of gravity looks very much like the electrostatic force between charged particles. Both diminish as the inverse-square of the distance between the bodies carrying electric charge or mass. It was the second principle that was the deal-breaker for many physicists, yet the Bohr model was tentatively accepted by physicists for the simple reason that it worked. It explained a whole raft of experimental results gleaned from the study of how hydrogen atoms emit light, which I'm going to discuss in more detail in the next chapter.

With Bohr's model, physicists could at last make accurate mathematical predictions of the wavelengths of light emitted by hydrogen atoms, which no other model had been able to accomplish. It was truly a devil's bargain. If you allowed Bohr's bizarre postulate about electrons not emitting energy in stable orbits to stand, you could explain the hydrogen atom. Without this postulate, you had no explanation for why hydrogen atoms only emitted light at specific wavelengths and no others. So as had happened many times in the past, a radical and even counter-intuitive idea had allowed progress to be made in making mathematical predictions about how matter and forces interacted. After further experimental tests, the idea was accepted at face value and added to a growing catalogue of 'Laws of Nature'.

The pace of discovery now increased at a break-neck speed. No sooner had Rutherford's model been publicized in 1911 and Bohr's re-vision of it appeared in 1913, than in 1915 the Prussian physicist Arnold Sommerfeld added an important tweak to how the electron behaved. Bohr had forced electrons to move in only specific orbits so that their energies were represented by integer multiples of a basic atomic energy. That is what would lead to hydrogen atoms only emitting light energy at specific wavelengths. Sommerfeld added the idea that the momentum of these particles also had to be considered. Instead of the orbits being perfect circles, orbits that were elliptical could also arise. What this did was to make the momentum of the electron as it orbits yet another quantity that came in integer multiples, and was reflected in the shape of the orbit. The more momentum, the more elliptical the orbit. This was an almost exact reflection of how objects orbited the sun. Planets travelled in nearly-circular paths while comets took very elliptical paths.

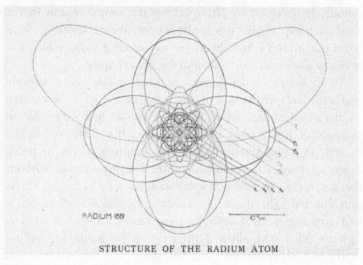

STRUCTURE OF THE RADIUM ATOM

The Bohr-Sommerfeld radium atom.

It had taken thousands of years for the subject to evolve from Aristotle's Elements to Dalton's atoms, and a century to cross the historical trajectory from Dalton's formless atoms to the Bohr model. In the years just before World War I, the world was in political tumult and about to embark on the War-to-End-All-Wars with Germany pitted against England. Meanwhile, physicists in these two countries within a matter of a decade had banded together through letters and journal articles to unravel one of the deepest mysteries in the world: what is the nature of matter? Through a series of brilliant experiments and insights, they had at last won their way to a model of the atom far beyond what Democritus or Dalton could have imagined. By weaving together seemingly disparate discoveries in chemistry, electricity and radioactivity, a new atomic theory had been fashioned for the 20th century. But as it would turn out, this was only the penultimate step in an even deeper series of discoveries to begin soon after World War I came to an end.

CHAPTER 3:

The Quantum World

If our knowledge of the atomic world had ended with the Bohr-Rutherford atom, believe it or not, we would still have all of the chemical compounds and drugs we enjoy today. Computers, radar, television, radio and many other technologies would still be with us, too. Through trial and error, we would even have discovered curious compounds called semiconductors leading to the invention of transistors. Even lasers would have been invented pretty much on schedule because they only depend on a knowledge of the atom with the sophistication of the Bohr model. But the whole point of scientific advancement is not to create new technological 'toys' by whatever means possible like alchemists stirring a pot, but to develop a deep understanding of why things are the way they are. What physical laws make semiconductors used in transistors possible? Why does radioactivity occur?

Exactly how does a single atom of hydrogen produce light? The next huge jump in knowledge about the nature of matter appeared, not through a better knowledge of chemistry, but from a completely different line of investigation.

Light as waves and corpuscles

Although the ancient Romans knew that prisms could separate sunlight into colours, in 1666, Sir Isaac Newton performed a number of careful experiments with light and prisms to deeply understand how light and colour were related. After a lot of effort, what he decided was that light was some kind of material substance that contained all of the colours as some property of its substance. This may seem like a let-down, but until Newton had studied colour in this way, it was common to believe that the colours of a rainbow or prism were produced by the medium that light passed through, or that the eye itself was creating the spread of colours. In fact we know from brain research that nature is actually completely colourless. It is our brains that create the impression of 'red' or 'green'. All nature gives us is a set of sterile wavelengths.

Around the same time that Newton was exploring prisms and the origin of colour in 1666, the idea that light was a wave phenomenon in what was called the *Luminiferous Ether* was proposed by Robert Hooke and Christiaan Huygens between 1665 and 1678. But Newton's experiments had led him to the competing idea announced in his 1675 book *Hypothesis of Light*, that light consisted not of waves but of particles called corpuscles. So by the start of the 19th century one hundred years later, scientists had to deal with the two competing schools of thought. The Wave Theory of Light explained some experimental results and the Corpuscle Theory of Light explained some others. The first scientist to actually measure the wavelength of visible light was Thomas Young in 1802 using a simple diffraction experiment involving light passing through a slit and falling on a screen. His answer was $1/43,636$ inches, or in modern measure

582 nanometres – right smack in the middle of the visible spectrum. That pretty much settled the debate because particles were discrete things and could not display the property of a wave, and so it seemed that Young's Wave Theory had won.

Isaac Newton's prism experiment.

Spectroscopy and atomic spectra

Meanwhile, as astronomers clamoured for larger and more distortion-free views of celestial objects, opticians such as Joseph von Fraunhofer began to develop high-precision optical systems that gave near-perfect images of stars and planets. But Fraunhofer realized early on that ordinary sunlight was too imprecise to serve as a guide for making precise lenses. He needed a source of light that was pure and as close to having a single wavelength as possible. This led him to create a device that took sunlight through a narrow slit and focused it in a prismatic grating to create the familiar rainbow. He then isolated the specific light he wanted to use to test his optical surfaces, and let this light pass further into his measuring instruments. A benefit of the slit was that it focused the spectrum in a way that a slitless rainbow or prism could not. What he noticed was that the rainbow spectrum now had dozens of dark lines passing vertically through the spectrum. His careful measurements of the wavelengths of 570 of these curious features, to this day called the Fraunhofer Lines, was quite an accomplishment, especially considering that he was otherwise rather uninterested in these lines. But his catalogue provided the grist for the mill for other scientists to investigate these lines further.

THOMAS YOUNG

Thomas Young (1773–1829) was born in Somerset, England, and was the eldest of 10 children. Even at an early age he was notorious for his precocious genius, allegedly having mastered Greek and Latin by the age of 14, but also acquainted with German, Hebrew, Aramaic, Arabic and Persian. He began his formal studies in medicine at the University of Edinburgh in 1794.

Thanks to a large inheritance, he became independently wealthy and by 1799 had established himself as a London physician, and by 1801 was appointed to the position of Professor of Natural Philosophy at the Royal Institution. His work on the nature of light between 1803 and 1807 before returning to medical practice became the seminal experiments that proved conclusively the wave theory of light. In particular his interference experiments using water in a 'ripple tank' are even today the classic experiments performed by every secondary school student in physics.

As a polymath, his interests also spanned many other fields of knowledge including the elastic properties of matter, the capillary action of fluids and even the translation of ancient Egyptian hieroglyphs. In the latter subject, in 1814 he had translated part of the text on the Rosetta Stone, which formed the basis for later work by Jean-François Champollion. Although Young married Eliza Maxwell in 1804 they had no children. Young died in 1829 at the age of 56 having suffered from acute asthma all his life.

Thomas Young.

Between 1820 and 1850, a bevy of investigators such as Gustav Kirchhoff, Robert Bunsen and Charles Wheatstone made a number of discoveries about Fraunhofer's curious lines using candle flames and materials burned in the flames. What became increasingly clear was that every element had its own particular set of lines with no known duplications. In fact, if an atom produced a dark line (called an absorption line) at a particular wavelength, it would also produce a bright line (called an emission line) at exactly the same wavelength. By the 1860s, William and Marguerite Huggins used these atomic fingerprint lines to identify the chemical composition of a variety of stars and nebulae. Something that philosophers such as Auguste Comte had proclaimed would be impossible to know only 25 years earlier.

Once again as for chemistry, it wasn't important to know how atoms produced these lines in order to apply this knowledge in identifying gases thousands of light years from earth. But for the sake of understanding how matter produced light, the presence of these lines provided an invaluable clue, but like determined crime detectives, scientists would have to dig hard to uncover what the clue was trying to tell them.

Atomic structure and quantization

Careful studies of the light from hydrogen gas by the Swedish physicist Anders Angström revealed that this light always appeared in the form of at least four bright lines at wavelengths of 410, 434, 486 and 656 nm. The line at 656 nm was especially intense and gave incandescent hydrogen gas its dramatic red glow. By some unknown process occurring inside an atom, there was a means for emitting light at fixed wavelengths rather than as a pure-white beam of energy spanning many wavelengths. The basic idea that you could stimulate an atom to give off light of a very pure colour was, by the way, not lost on store owners and advertisers (see box).

New laws of nature sometimes reveal themselves in patterns found in measurements. For instance, if you plot the height of the sun above the horizon for a full year, you discover the correlation of seasonal changes with solar elevation angle. The pattern of spectral lines for some elements seems rather haphazard, but for hydrogen there was a tantalizing pattern of spacing changes that literally begged for someone to sit down and work out the numerical pattern. Within a few years of the hydrogen spectrum being observed and published by Angström, the Swiss mathematician Johann Balmer in 1885 came up with a formula that predicted the wavelengths of these lines very accurately:

$$\lambda = 364.6 \; \frac{m^2}{m^2 - n^2}$$

where m and n were integers. To get the hydrogen lines, select n = 2 and m = 3,4,5 and 6, which is called the Balmer series. For example, with m = 3, the predicted wavelength is 656.3 nm, which corresponds to the very red hydrogen 'Balmer-alpha' line measured by Angström at 656 nm. But what did this numerical curiosity mean in terms of atomic structure? The answer to this puzzle required an entirely new perspective on the nature of light itself, and this perspective came about by studying a very common phenomenon. If a blacksmith takes a bar of solid iron and heats it, it will start to glow as it warms up. The sequence of colour changes is not random but follows the exact sequence every time starting from a dull red glow, through orange and yellow, to white. You can perform the same experiment using the hot plate on your stove.

NEON LIGHTS

The French engineer Georges Claude was well-versed in the science of electrical discharge tubes, which were studied during the later decades of the 19th century. There was also an abundant supply of neon gas produced by the liquefaction of air to extract hydrogen and oxygen. Putting two-and-two together, Claude developed neon discharge tubes and on 18 December 1910 he rolled out the first neon sign for the Paris Motor Show. It was an instant hit. This invention revolutionized how businesses captured the

Writer's Block Bookstore
32 W. Plant Street
Winter Garden, FL 34787
(407) 335-4192
5/20/2022 5:10:31 PM
REG #: 5   CLERK #: 114   TRAN #: 33819

1@	$12.99 9781398802346	$12.99

Quantum Physics: From Schrödinger's Cat

1@	$12.99 9780593203392	$12.99

Frankenstein

1@	$13.99 9780593203385	$13.99

Dracula

Sub-Total: $39.97
Tax $2.59
Total: $42.56
Tendered: VISA $42.56
XXXXXXXXXXXXX0163 mm/yy APPROVAL 047824
Transaction ID: VISA
Reference No: 31474
Transaction Type: Sale

interests of prospective patrons, and most downtown areas of major world cities have huge, even garish and moving, electric light displays based on Claude's clever invention. Next to Edison's electric lights and Fleming's vacuum tube 'valve', neon signs were the first dramatic, easy to understand, application of our growing knowledge of the quantum world.

The quantization of light

Well, scientists were not content just to note how colour changed with temperature, so during the 19th century they perfected instruments that were capable of measuring light intensities of objects emitting these different colours under temperature-controlled conditions. The light intensity curve didn't just have a fixed shape for one temperature, this shape remained the same even as the temperature increased. The curve had a single peak where the light intensity would be the brightest, and this peak shifted to shorter wavelengths and also increased in strength as the temperature was increased. The relationship between temperature and light caused some physicists such as the Austrian physicist Ludwig Boltzmann to explain the relationship using the new science of thermodynamics, but the predicted shape didn't look at all like the actual intensity curve. It wasn't until 1900 that the German physicist Max Planck made a change in how the curve was calculated by introducing a mathematical 'trick'.

If you assumed that light came in packets of energy rather than a continuous wave, the calculation of the light intensity curve matched exactly what was observed. The curve, called the black body curve, also had the property that its peak would shift to shorter wavelengths as a body was heated, just as had been observed. Planck's method required that the light packets he called quanta, came in fixed energy units given by a formula $E = h\nu$, where the Greek letter ν is the frequency of the light and h is a constant, which today is called Planck's constant. For example, light in the ultraviolet part of the visible spectrum has a frequency of $v = 800$ trillion cycles-per-second (called a hertz or Hz), and with $h = 6.6 \times 10^{-34}$ joules-Hz, you get an energy for each ultraviolet light quantum of $E = 5.3 \times 10^{-19}$ joules. This doesn't sound like much energy at all, but these quanta can actually penetrate cells and damage DNA molecules causing mutations or cell death! The light quantum is such a fundamental particle that it was re-named the *photon*, but not until the American chemist Gilbert Lewis used this term in an article in the journal *Nature* on 18 December 1926. Physicist Arthur Compton then started using this name in 1928 and it finally was adopted by physicists from then on.

Understandably, the idea that light came in tiny energy quanta didn't make sense to many physicists. Maxwell had just developed an entire theory of light as a continuous, wave-like phenomenon, and all of the optical experiments with reflection, refraction and diffraction proved pretty convincingly that light was a wave phenomenon. Even Planck himself didn't like his own idea and thought it was just a convenient mathematical trick to derive the correct intensity curve. It was to him very much like the meat stuffing in a frankfurter. There isn't anything in the meat that demanded it can only be packed into frankfurters of a specific length. But once this quantum idea circulated in the physics community, a young patent clerk named Albert Einstein in 1905 took this same idea and used it to explain how some metals can eject photons when struck by light.

In modern-day photography, the idea of the photon is fundamental to how your smartphone digital cameras work under low-light conditions. You can see rather dramatically how light comes in quanta as you let your camera use shorter and shorter exposure times, capturing fewer and fewer light quanta to build up the photographic image. The image gets grainier and grainier until it becomes lost in the random noise of the imaging chip. A precise demonstration of this can be performed with sophisticated equipment and shows the dramatic process step-by-step as exposure times are increased.

THE PHOTOELECTRIC EFFECT

The 'photoelectric effect' had been known since 1887 when Heinrich Hertz discovered it while experimenting with radio waves and electrical sparks, but no one could figure out why it existed using Maxwell's wave theory of light. The problem was that, when you increased the intensity of the light, more electrical current did not flow, but if you increased the frequency of the light above some threshold, you could in fact get more current to flow. Somehow, it wasn't just the wholesale rush of energy by all of the light rays of different frequencies that caused the effect, but the effect only occurred for light waves above a specific frequency.

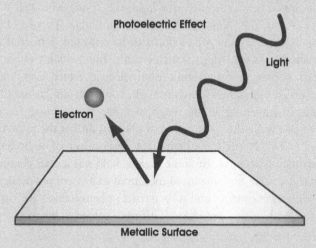

Photoelectric Effect

Light

Electron

Metallic Surface

The photoelectric effect is what lets solar panels generate electricity from sunlight.

This relationship is exactly what Einstein predicted by using Planck's light quantum idea. So, instead of the light quantum being merely a mathematical trick, it was a real phenomenon of light that could be independently measured. There were now two very different ways to use Planck's energy quantum relationship and to measure the value of Planck's constant using entirely different phenomena. The photoelectric effect was so important in proving Planck's light quantum idea that Einstein won the 1921 Nobel Prize for this work, and not ironically for his Theory of Relativity.

The particle-wave duality of light

But now things get really, really strange. Remember we have two competing ways to describe light: the Young-Maxwell Wave Theory and the Newton-Planck Corpuscular Theory. Light really does behave as waves that can be reflected, refracted and diffracted in a way that particles can't. But Planck's curve for heated bodies and Einstein's photoelectric effect also loudly proclaimed that light is a quantum phenomenon similar to bullets of energy travelling at the speed of light. You cannot describe Maxwell's wave theory in terms of photons and at the same time preserve his detailed mathematical description of light as an electromagnetic wave. By some means, light has a dual character in that it can be described and measured as a wave phenomenon for some experiments and as a particle phenomenon for others but not both at the same time. This may not be as bizarre as you might imagine!

Water can behave as a wave that produces ripples on the surface of a harbour, and can interfere with itself as troughs and crests happen to be at the same spot on the water's surface. The results can be a beautiful and complex patina of ripples. Under different conditions, water can behave as quantum particles called hailstones, but to describe the formation and trajectories of hailstones we cannot use waves. In essence, depending on which experiment you choose, the water will behave either as a particle or a wave but not both. For light, this relationship is even more intimate than harbour waves and hailstones.

Every secondary school science class studies water waves in a tank, where a single slit or opening allows water waves to travel from one side of a barrier to the other. Although the waves are completely parallel to the wall on one side, after they pass through the slit they bend into beautiful semi-circular arcs. When two slits are added to the barrier, the crests and troughs from each arc interfere with each other to produce a more complicated, but still very regular, pattern. This simple experiment was performed by Thomas Young in the early 1800s to demonstrate convincingly how light interference works according to the wave theory. It is widely regarded as one of the most beautiful and important experiments in modern physics.

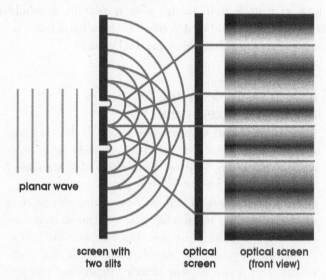

planar wave

screen with two slits   optical screen   optical screen (front view)

Interfering water waves in a two-slit experiment showing wave interference.

The discovery of the light quantum also had a direct implication for understanding internal atomic structure. Light behaves as a quantum particle that carries a specific kernel of energy that depends on its frequency according to Planck's Law $E = h\nu$. This turned out to be the missing ingredient that explained Bohr's atom. Each of the orbits the electron could take was defined in such a way that when the electron jumped from one orbit to another, it either absorbed or emitted a photon of light energy with a specific frequency set by Planck's Law. This could only mean that the orbit energies *were also quantized* so that they gave exactly integer numbers of photons during the light emission process that produced the Balmer lines of hydrogen and at no other frequencies. So not only did Planck's light quantum hypothesis explain the Planck curve and the photoelectric effect, but it also explained how hydrogen and other atoms emitted light at only specific frequencies. But there was one final, bizarre, phenomenon that became the keystone to an entirely new way to look at matter itself. In this next revolution, it would not only be light that could be described as either particles or waves. Electrons themselves had this same dual nature!

HOW TO PERFORM YOUR OWN DOUBLE-SLIT EXPERIMENT

There is an equivalent experiment to the double-slit water wave demonstration that involves only light of a single wavelength. Today, anyone can perform this experiment using a toy laser pointer. All you need to do is to take a piece of aluminium foil and cut two narrow slits as close to each other as possible with a razor blade making sure the slits are parallel. Then shine the wide beam of a laser pointer at these slits. On the other side of the foil about 10 cm away, place a white sheet of paper, and you will see the interference pattern resembling what you find with water wave interference. Congratulations. You have just re-performed Thomas Young's experiment from 1802! But now there is the new 20th century trick that will reveal how light is quantized using only a good-quality digital camera, or even a smartphone!

Replace the white sheet of paper with the imaging chip of a digital camera just behind the slits at about 10 cm. Now, instead of sending the full avalanche of photons through the slits, turn down the wattage of the laser and your exposure speed so that your camera only captures one photon at a time going through the slits. At first the images will show random dots where the photons hit the imager, but after you record 100, 1,000 or 10,000 of these events you will once again see the same double-slit pattern you saw for wave interference.

When water waves passed through the system between the slit barrier wall and the display screen, the waves interfered with each other to produce the pattern you see on the screen. But when you send one photon at a time through the slits, miraculously they act as though each photon has actually taken paths through both slits at the same time and interfered with itself – otherwise its spot on the screen would not be located in such a way that it would add up to the wave-like interference pattern. This is such a bizarre and confounding situation that we will return to it in a later chapter.

CHAPTER 4:

Things that Wave and Things that Don't

Quantum mechanics deals with special kinds of waves in nature. With the exception of ocean waves or sound waves, we tend not to be familiar with things in our world that have to do with waves. Some kinds of waves called 'travelling waves' move through space like the ripples on a pond from a dropped stone. Other kinds of waves remain frozen in space. It is these 'standing waves' that play a crucial role in understanding quantum mechanics. They are hard to find in nature but, once you know what to look for, they can be found almost everywhere, and are as accessible as the strings on a piano, a violin or a guitar.

The familiar media of water and sound waves

The easiest kind of wave, and the most fun, is the water wave produced by a stone dropped into a quiet pool of water, or a dramatic breaker you can ride on your next beach holiday. It is a vertical displacement of water that is controlled by two forces: water tension and gravity. Gravity pulls the surface downwards towards the centre of the earth, while water tension places a limit on just how high it can rebound above the flat undisturbed water level. Thanks to a number of laws like the conservation of mass, momentum and energy, this motion has to be periodic in time as the wave moves from its origin to its endpoint on the beach. The height is also determined by the local depth of the water basin, or the continental shelf that defines the slope of the beach. When far away with a very deep basin, you get ocean waves only a few centimetres high, but when they arrive at the beach, the conservation principles and the shallowness of the beach cause the wave heights to increase to several metres or more. For surfable breakers this steepening can be a lot of fun if no sharks are around, but for tidal waves this effect is almost always deadly.

When the medium is water with a fixed two-dimensional surface you get water waves. For sound waves in air or in a solid body, however, there is no fixed two-dimensional surface because the phenomenon is three-dimensional. Sound waves are caused by the change in density of a compressible material travelling at the speed of sound in the medium. This medium can act like a bouncing spring as the external force is met by the intermolecular resistance of the medium and oscillates back and forth to strike a balance. For air at sea level, the speed of these pressure waves can be 1,000 km/h. In a solid body such as the earth's crust, the speed can be 20,000 km/h and these are called P-waves (e.g. pressure waves) by seismologists. Instead of the one-dimensional height of a point on a surface above some reference level changing in time, as in the case of water waves, for pressure waves it is the local density of the air or rock that changes in time. But not all waves actually move through space.

Air can act like water to produce waves in the atmosphere that actually do not move. These are called gravitational standing waves and are often seen when cold air is pushed up and over the top of a mountain chain. When it comes down on the other side, this heavy air falls under its own weight. As the cold air mass travels away from the mountain it bounces up and down under gravity forces from its weight, and atmospheric compressional forces. As the air reaches higher altitudes the water vapour can condense into clouds leaving behind what look like waves in the atmosphere. The wavelengths are determined by the speed of the cold air and the specific geometry of the mountain range.

So far we have considered air, water and rock as media that can 'wave' and produce a variety of wave-like phenomena. Now we are going to venture into an arena of things that are far less

Atmospheric gravity waves and clouds produced by vapour condensation.

familiar. When we considered water, rock and air, these are media that themselves have an obvious solidity. Even without the 'wave' you can hold water, air and rock in your hands... well for air at least you can do so in principle! But for the last 200 years scientists have known about a collection of things that are invisible and cannot be touched, yet they can be measured. These are the electric and magnetic fields that emanate from every charged object and every magnet. We are pretty familiar with what magnetic fields look like thanks to secondary school experiments with magnets and iron filings. Electric fields are a bit less obvious and extend outwards from the surface of a charged body and are often drawn like the spokes on an invisible wheel. There are no limits to the extent of these fields because they travel outwards from each object at the speed of light starting at the instant the charge or magnet is created. To see how this works, let's look at electric fields.

Waves in unfamiliar media

If you start with a charged body at rest, its electric field decreases in strength away from the charged body into the surrounding space. If the particle was created one second ago, the outer edge of the surface of this spherical electric field is one-light-second away; some 300,000 km. Not to labour the obvious, but before the charge came into being, there was no electric field. The instant after it came into being its electric field extended out into space a distance equal to the time it has been in existence multiplied by the speed of light. You may not have instruments sensitive enough to be able to measure it at its farthest limit, but it is there nonetheless.

Now, if you move this charged body one metre to the right, its electric field 'spokes' will be centred on that new spot and also travelling outwards at the speed of light. What happens is that a kink has developed in the electric field. Inside it is pointed in one direction, and outside it is pointed in a slightly different direction. Within the kink, the electric field has to change from one to the other directions. Now if you move the charge back and forth between the two locations one metre apart, you will create an electrical kink disturbance with a wavelength of 1 m travelling outwards at the speed of light. This is called an electric wave and a similar phenomenon is what allows telegraph systems to work when the electric field is switched on and off with each key motion. The electric wave travels through the telegraph wires at the speed of light and causes a distant telegraph key to also change from on to off as the electrical field changes at the receiving end. Now let's look at magnetic fields.

We have all seen the lines of magnetism around a toy bar magnet looping out into the space and then returning at the other pole of the magnet. Michael Faraday was so impressed by these lines, first studied by the philosopher René Descartes in 1640, that he considered these 'lines of force' to be the foundational element of all things in nature, even gravity. But these magnetic lines are not fixed in space like the tracks travelled by trains between different towns. In fact, if a magnet were placed in a gas of charged particles called a plasma, these lines can move about as the plasma shifts position. Under certain conditions, lines of magnetism can produce waves that travel down the lines at a speed set by the strength of the magnetic field and the density and charge of the surrounding plasma. In 1942, the Swedish physicist Hannes Alfvén was the first to propose that these waves could exist, but it took decades for them to be spotted in the magnetic environment of the sun's surface and corona. Alfvén waves actually seem to transport energy into the corona of the sun to keep it heated to millions of degrees Celsius.

MICHAEL FARADAY

Michael Faraday (1791–1867) was born in Surrey, England, to an impoverished family of four children to a blacksmith. Lacking much formal education, Michael had to teach himself and eventually became the apprentice to a bookbinder in London. He read voraciously and became interested in science and especially the growing subject of electricity.

By the age of 20, he was sitting in on lectures by Humphry Davy of the Royal Institution and sent Davy a 300-page tome containing his lecture notes. Davy was so impressed by this that, following a disabling accident that blinded him, Davy appointed Faraday as Chemical Assistant in his lab. Following his marriage to Sarah Barnard in 1813, in his later years he became the Deacon at his church where 'a strong sense of the unity of God and nature pervaded Faraday's life and work' in the words of science writer Jim Baggot. During his time at Davy's laboratory, Faraday worked with Davy to invent the first electric motor in 1821, and soon after his

Michael Faraday.

interests turned to electromagnetism after the death of Davy in 1832.

Through a series of notebooks, his struggles with magnetic fields and currents could be followed step by step until his pioneering discovery of electromagnetic induction, which set the stage for the invention of the electric dynamo. Faraday's great theoretical contribution involved his proposal that fields of force govern natural phenomena. His concept of lines-of-magnetic-force permeating space was central to the work by his successor James Clerk Maxwell. He was also active in investigating industrial pollution.

Beginning in 1827 he gave a series of public lectures on science that became legendary for their humour and dramatic demonstrations of a variety of physical phenomena in chemistry and physics. These lectures continue to the present day in his honour as an educator.

One of the most peculiar, and at the same time well-known, waves is light itself. For centuries it was thought that it needed a medium called the *Luminiferous Ether*, or simply 'Ether' for short, in order for it to travel through the vacuum of space like an ocean or a sound wave, but Maxwell showed how this was not at all necessary. In an earlier example I described electric waves produced by kinks in the electric field from a moving charged particle. According to Maxwell's mathematics, the kink in the electric field does not travel alone. Simultaneously it generates a magnetic field within the kink because, as Hans Ørsted had discovered back in 1820, a moving charge or current produces a magnetic field. For our electric field kink, there will also be a magnetic field kink that will vary in strength perpendicular to both the direction of the electric field and the direction of travel. This electromagnetic disturbance travels at the speed of light and its properties can be described in terms of the wavelength (or frequency) of the charge's position change. The reason that light can travel through an empty vacuum is that the vacuum is not actually empty. It contains the electric field of the charged particles, and this field is what transports electromagnetic disturbances through space from point to point. Now we come to a new kind of wave that is positively amazing.

Waves in the strangest medium of all

Contrary to what you have been taught in school, gravity is not a force in the same way as the other forces in nature such as electromagnetism. Although as Newton showed, gravity produces an attraction between two bodies that decreases by the inverse-square law, the origin of this effect is very different. Imagine a cake covered with frosting. The frosting represents the various fields in space that causes forces such as electric and magnetic attraction or repulsion. The body of the cake represents three-dimensional space and what we call the gravitational field. The gravitational force is measured by how much the geometry of space is distorted. In fact, the effect of gravity can be thought of as purely a geometric distortion without any mention of an external force to produce it. One of the predictions of this interpretation of gravity developed by Albert Einstein in his 1915 General Theory of Relativity, was that these geometric distortions can move as waves from one place to another. What causes these waves is the act of accelerating an object. The stronger the acceleration, the more intense the distortion wave that travels from the object to other locations in space. The way you detect these distortions is very simple: you just measure very accurately the distance between two or more objects. When one of these gravity waves passes by, it will cause the distance between two objects near you to change in time, first making them slightly closer and then slightly farther away. But this wave is carried by the 'medium' provided by the geometry of space, more correctly called spacetime in relativity theory. What exactly does it mean for the geometry of space to be its own medium? That is where

Einstein's theory of general relativity departs from the classical Newtonian model for space and time. It is, conceptually, one of the most challenging ideas to understand even though the mathematics are quite clear about the interpretation.

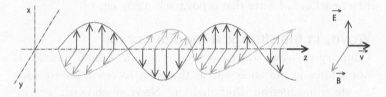

In an electromagnetic wave, the changing electric field produces a changing magnetic field and both travel through space on two different perpendicular planes.

Neither magnetic fields nor electric fields distort the space in which they are embedded. If you used a light ray to measure the distance between pairs of points in three-dimensional space within an electric field, those pairs of points would all yield the same value for the length of a metre stick. For gravitational fields, a difference does occur, which is interpreted by local observers to be a gravitational 'force' causing acceleration. This geometric distortion of three-dimensional space was explored in great mathematical detail by Carl Friedrich Gauss and Georg Bernhard Riemann in the early 1800s, leading to an exhaustive description of 'N-dimensional, non-Euclidean geometry'. The main ingredient was a quantity called the metric tensor represented in relativity theory by the symbol $g_{\mu\nu}$. All of the information about how a particular space was distorted was to be found in $g_{\mu\nu}$. To reveal what paths objects travelled in these spaces, all you had to do was to use Riemann's mathematical tools to uncover their mathematical properties, such as what the straightest-possible line (called a geodesic curve) would look like.

What Albert Einstein proposed from his ideas about relativity was that the only thing you could measure or have access to were the paths of particles called worldlines. The relativistic information coded in these paths defined what the geometry of space, now called spacetime, would be, but spacetime *did not pre-exist* the worldlines. When he worked out the mathematics needed to describe this relativistic idea where the worldlines were the foundational element of the physics, he re-discovered the work by Riemann and how the metric tensor $g_{\mu\nu}$ was of key importance. In essence, he worked backwards from knowing the geodesic curve for a particle to what the metric tensor had to be. But for Einstein, this metric tensor was nothing more than a statement of what the gravitational field would look like that was consistent with the geometric distortions experienced by worldlines of particles. To Einstein, *first* you had worldlines, *then* you had spacetime not the other way around as was the case for Riemann's mathematics. This is why Einstein's theory of gravity is called background-independent, which is an idea we will get to in Chapter 13. Another way to look at this is that $g_{\mu\nu}$ representing the geometry of spacetime is identical to the $g_{\mu\nu}$ that represents what we like to call the gravitational field.

CONFIRMING EINSTEIN'S THEORIES

The theoretical prediction of gravitational waves by Einstein was finally verified nearly 100 years later in 2012 when two massive black holes just over one billion light years from earth collided with each other and produced a huge ripple in space. After travelling across the universe for 1.4 billion years, this geometric distortion wave finally reached earth on 14 September 2015 at 09:50:45 UT, and was detected by the advanced LIGO/VIRGO gravity wave telescopes in Hanford, Washington, and Livingston, Louisiana. Gravity waves travel through the pure vacuum of empty space and do not require any other fields to support them. They are a feature of the very geometry of our three-dimensional world.

So, this little inventory shows things that wave in some kind of medium be it air, water, rock, electric or magnetic fields. Even the geometry of space itself can serve as a medium for gravity waves. All of these waves have the common feature that they travel through space and change in time. Also, all of the media through which the waves travel are physical things, yet some kinds of physical waves do not even move at all. When you pluck the string on a guitar, the string vibrates with a wave but the wave is fixed between the ends of the string and form what physicists call a standing wave. Other examples of standing waves are the pressure changes in the pipe from a church organ, and the light wave in a ruby rod for a common laser. These are still all familiar because they involve physical media like wire, air and electromagnetic fields. In the next chapter, physicists were introduced to a new kind of wave that, quite frankly, is still mysterious to physicists even in the 21st century.

The Virgo telescope detects gravity waves by measuring the distances between mirrors many kilometres away.

CHAPTER 5:

Matter Waves

The wave-like properties of light can be traced back to elementary physics and the behaviour of the electromagnetic field. The wave-like behaviour of material particles such as electrons and even entire molecules has no equivalent source. Matter waves are the keystone phenomenon that allows quantum mechanics to exist and to make astonishingly accurate predictions about atomic physics and the interaction of light with matter. It is also the basis for even more advanced theories about the nature of the physical world and the processes of observation and measurement. The proof that matter waves exist is extremely simple. Like photons of light, individual electrons can pass through a double-slit experiment and apparently travel through one or the other slit, but when thousands of these individual encounters are combined, the resulting collective behaviour is that of wave interference just as for light.

The journey towards an understanding of quantum mechanics

You may not realize this, but we are actually climbing a mountain. Every now and then we reach a temporary resting spot and can look out at the surrounding vista, but as we climb the effort gets a bit harder with every mile. Also, the vegetation changes from a lush forest with familiar waterfalls to a sparser forest with trees and shrubbery we have never seen before and for which we have no familiar names. Further ahead, we seem to catch glimpses of bare rock and granite and perhaps a snow-capped peak as what we hope are our ultimate destinations. Our botanist and geologist friends have names for all these different things that are new to you, but because you were not trained in their language you can't follow along in their excitement very easily. What kind of magnificent view we will be treated to closer to the summit we can only dimly imagine, but we have convinced ourselves that what we may see will be worth all this arduous effort. We trust the signage at the start of this climb that promises us an incredible view.

In learning about the nature of matter we have embarked on a similar climb. Our guides tell us to hold on for just another few kilometres because what we will find near the summit will literally blow our minds. Along the way they tell us about the different ideas we are encountering so that when we finally do get to the summit, we will have a better idea of how we got there and by what logical paths. The best guides will give us many analogies and familiar terms to hang on to like safety ropes dangling from the mountainside. But what the guides don't tell us is that they will frequently have to add cryptic phrases and strange names to the conversation. Although the guides try, helpfully, to provide definitions and draw pictures on paper to show what is going on, even the definitions can sometimes be hard to understand in themselves. This is how it has always been for much of human history, where around countless campfires shamans and mystics have tried to tell the *Story of the World* with the best tools and terminology at their disposal. The singular difference between then and now is that the scientific story is one of hard data, experimentation, careful observation and the deep logical connective tissue provided by mathematics. More often than not, it has been the mathematics that has led us to deep insights, and so our story-telling borrows heavily from mathematical statements. In few other areas in physical science does this ring as true than in the study of the nature of matter beyond the atomic realm. We form mathematical models of what we are observing based on logical and self-consistent patterns in our data, and in some sense actually believe these mathematical models are a direct image of real things in nature.

The basis for matter waves

By 1923, the physics community was still abuzz with the idea that light could be a wave phenomenon described by its relationship $c = \nu \lambda$, with c being the speed of light, λ being its wavelength and ν being its frequency, but could also behave as a particle described by Planck's light quantum relationship $E = h\nu$. Albert Einstein, meanwhile, showed from his relativity theory in 1905 that in addition to his iconic formula $E = mc^2$, there was also a relationship for light such that $E = p/c$ where p is the momentum carried by the photon. These four simple equations essentially told us all we needed to know about the wave and particle properties of light. In fact, if you equated $E = h\nu$ with $E = p/c$ and used $\nu = c/\lambda$, you got a new relationship for light $p = h/\lambda$ that summarized its particle and wave properties. This insight was so deep and elegantly stated that it would easily fit on a bumper sticker or a T-shirt. Meanwhile, this light quantum idea had been used to improve the justification for Bohr's atomic model and how electrons within atoms only emit specific wavelengths of light. To recap, by the second decade of the 20th century we have light being represented as a photon or an electromagnetic wave, and the object that emits light such as an electron, being solely described as a particle. One might ask, why isn't there a better symmetry between these descriptions so that both electrons and light could be described as either particles or waves? This was exactly the question that Prince Louis de Broglie asked in his PhD thesis in 1924.

What de Broglie did was simply to look at the relationship $p=h/\lambda$ and turn it around so that $\lambda=h/p$. This simple algebraic operation could be performed by any secondary school student, but in de Broglie's hands it meant something quite profound. What this now says is that the wavelength of a photon, λ, is related to its momentum, p. But momentum is a feature of material particles, too. A mosquito, baseball or planet all have their own momenta, which is nothing more than just the product of their mass and velocity, $p = mv$. So, de Broglie proposed a new idea by writing down the formula

$$\lambda = \frac{h}{mv}$$

In no more than four familiar symbols suitably arranged, he logically showed that matter, m, had a calculable wave-like character, λ. This logical conclusion was also the last ingredient for explaining Bohr's atomic model with its restriction that electrons could only be found in specific orbits and no others. In the last chapter, I described standing waves, which are waves that do not move. According to de Broglie's new understanding, Bohr's electron orbits were just places inside the atom where the circular, electron matter standing wave could exist without destroying itself.

LOUIS DE BROGLIE
Louis Victor Pierre Raymond de Broglie (1892–1987) was a member of the aristocratic family of Broglie. Although the family was known for the many military and political posts they attained, Louis was an anomaly. He had planned to

pursue a career in the humanities and political history, but after receiving his first degree in history his interests took him to mathematics and physics.

During World War I he worked on a variety of radio communications problems for the French Wireless Communications Service stationed at the Eiffel Tower. After the war, he resumed his work on a PhD in physics and in 1924 presented his thesis 'Research on the Theory of the Quanta', which would later set the stage for improving the early quantum theory of atomic structure. Much of his early work was conducted at the laboratory of his older brother Maurice where he studied the absorption and emission of x-rays by atoms. In 1929 he received the Nobel Prize in Physics for his discovery of the wave properties of matter.

Louis de Broglie.

With light, its nature as a wave can be described in great detail as a disturbance in the electric field of a moving charged particle. The electric field is a physical thing that provides the medium for sustaining wave motion. But de Broglie's matter waves are very different. They are not the same thing as the waves in the electric field of the particle but are an entirely separate physical phenomenon. We envision the electron for the definite particle that it is, but what is this thing associated with the electron that is measured in terms of a wavelength? What is it that is waving?

What is it that is waving?

In science, particularly physics, it is often the case that we use mathematics and mathematical models as a means to form an image of what an object or phenomenon is. Our mathematics serves as a microscope or a telescope to collect information and observations into what we hope is a logically consistent picture of what we are 'observing' in domains far too vast or too small for us to directly observe them. We create mathematical models of the interior of earth using seismic waves to discern the various rock layers that can never be directly studied by actually visiting them. With artistry, the models can be turned into pictures of what we would see if we were there, entombed no doubt in solid rock. When we try to use our mathematics in this way to create a consistent image of the electron, the process fails.

In our mind's eye, we envision an elementary particle such as an electron as a small sphere with a mass of 9.11×10^{-31} kg. On its surface, it carries 1.6×10^{-19} coulombs of negative electric charge. Its diameter, however, is undefined because no measurement ever conducted to measure its diameter actually detects a physical surface. All we get from our measurements are upper limits to how 'big' the electron could be without its surface actually being detected. Currently this limit is at 10^{-18} m, which is more than 100 million times smaller than the diameter of a hydrogen atom. In fact, most of our advanced theories, including Einstein's relativity, require the electron to be quite literally an infinitesimally small point in space. What this means is that an electron cannot be a tiny sphere of matter but has to be a vanishingly small 'something' that behaves as if it has a specific quantity of mass and charge. On the other hand, de Broglie's matter wave idea says that in some sense an electron is a wave-like object that like all waves can be extended in space and interfere with itself on the scale of an atom. This is exactly the opposite of it having a vanishingly small size and is the deep paradox encountered when thinking about matter as both a particle and a wave. Prime Minister Winston Churchill remarked about Russia in a 1939 BBC interview that 'it is a riddle, wrapped in a mystery, inside an enigma'. A similar sentiment can be offered for the nature of electrons, which seem to be a collection of conflicting information and mutually exclusive logical deductions. To get around this issue of electrons as waves, de Broglie came up with a clever idea: Pilot waves.

You can well imagine that de Broglie's prediction that matter should have wave-like properties immediately got the physics community interested in testing this provocative prediction. One of the classic features of light as a wave is its ability to show the properties of wave interference and diffraction. This would be a technically difficult experiment to perform with electrons, but in 1927, two groups of experimenters proved this effect rather dramatically. At the University of Aberdeen, George Paget Thompson fired a beam of electrons at a thin film of celluloid and saw the characteristic diffraction effect of concentric circular rings, but formed from scattered electrons striking the photographic film rather than light waves. Meanwhile, between 1923 and 1927, Clinton Davisson and Lester Germer at Western Electric/Bell Labs used electrons passing through a thin film of nickel and saw the same effect. They, however, took the same nickel film and passed x-rays through it to confirm that the electrons and x-rays were experiencing the same diffraction effect. The spectacular side-by-side photographs soon appeared in every college textbook on atomic physics through the end of the 20th century. Both teams won the Nobel Prize in Physics in

1937 for having verified this amazing new property of matter; a property never considered in the millennia of investigations that came before.

PILOT WAVES

Imagine visiting Hawaii and watching the waves crash onto the beach. Almost always you will discover humans with their surfboards riding the crests of these waves with varying degrees of success. It is pretty obvious that the humans are of a different form and substance than the water wave they are riding, but the behaviour of the wave dictates entirely the behaviour of the surf-riding human particle. According to de Broglie, electrons are particles with a finite location in space, but they are also surfing a hidden phenomenon that he called the pilot wave. He had no idea what the substance of these pilot waves were, only that they dictated exactly how the electron would move and give rise to its associated wave-like attributes. Just as Sir Isaac Newton could never explain exactly what gravity was, in itself, he nonetheless went on to uncover a wealth of new physical phenomena

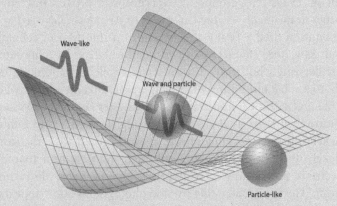

A hypothetical model of de Broglie's pilot waves steering an electron.

*with mathematical precision. Similarly, de Broglie set the
stage for a revolution in physics based on matter waves, and
still to this day we have not the slightest idea for matter
particles what it is that is actually 'waving'.*

In the intervening years between 1927 and today, ever-larger
collections of matter have been investigated to see if they, too,
reveal any hint of their wave-like properties. Matter wave
interference has been detected and even imaged in beams of
rubidium atoms, but why stop at mere atoms? In 2009, 60-
atom fullerene molecules were put through the diffraction test
and again showed matter wave interference. In 2011 the ante
was raised to 430-atom molecules by a team in Austria led by
Markus Arndt and Stefan Gerlich, again showing this effect.
By 2019, one of the largest molecules for which matter wave
interference has now been confirmed is the unpronounceable
molecule $C_{707}H_{260}F_{908}N_{16}S_{53}Zn_4$, which has over 40,000 protons,
neutrons and electrons. The effect was seen by the team at the
University of Vienna also led by Markus Arndt.

What we now had by the mid-1920s was nothing less than
a dramatically different physical world than what had been the
theoretical commonplace as recently as the turn of the 20th
century. After 100 years of atomic science based on the sanctity
of matter having a firm foundation, we now have a far more
insubstantial and paradoxical model of matter as simultaneously
a finite particle and also a wave-like phenomenon. Physicists
quickly turned their attention to exactly how wave-like matter
could emit wave-like light, or particle-like photons, depending
on which set of experiments you wanted to perform. It wasn't
enough to simply state that matter emitted light or even produced
an electric field. What was now under deep consideration was
the detailed process for how this happened, and for that matter
(pardon the pun) how it was that objects at the quantum scale

were actually observed and measured. Only at this scale could the curtain be drawn back on the very essence of matter as a physical thing fixed in space and time.

OBSERVING THE ATOM

Another recent vindication of the existence of matter waves was revealed in 2013 when physicist Aneta Stodolna and her team of physicists at the FOM Institute in Amsterdam succeeded in creating an actual image of the internal structure of hydrogen atoms using an instrument called a quantum microscope. What it showed was that electrons do indeed travel in specific paths around the nucleus, which supports both the Bohr atom idea and its basis in the existence of electron matter waves.

CHAPTER 6:

Quantum Waves

The wave function of a particle is a mathematical description related to the probability of finding the particle in a specific place at a specific time. For atoms, this function can be very complex because it changes depending on the energy and momentum of the electron. It is also complex because instead of being a point-like object, the electron behaves as a wave-like object capable of interfering with itself and its environment. Certain regions of atomic space are precluded because in these regions the electron's wave would interfere destructively with itself yielding a probability of zero, much like the vibrating strings on a guitar have nodes where the string is not in motion. Wave functions are not physical waves because they do not involve a vibrating medium to sustain them unlike water or sound waves. Instead, they act as though 'probability' is its own physical medium – a happenstance that is at the heart of a deep understanding of quantum mechanics itself.

So here we find ourselves sometime in the mid-1920s still learning more curious things about matter and the nature of atoms. Planck's light quantum idea applied to the way electrons emit and absorb energy led to an explanation for why Bohr's atom required electrons to have specific amounts of energy in integer steps; called energy quantization. De Broglie's matter wave idea, meanwhile, explained why electrons could only orbit at specific distances from the nucleus so that their waves did not destructively interfere. The revised Bohr-Sommerfeld model included the momentum an electron has in each orbit, which also came in integer units, called orbital angular momentum quantization. The combination of the Planck light quantum and de Broglie matter wave ideas had produced a completely self-consistent model for the atom that yielded the spectral lines observed by hydrogen, and any other atom that only had one electron. There was, however, one big problem: the periodic table of the elements is chock full of atoms with more than one electron. The Bohr-Sommerfeld model that worked so well for hydrogen worked increasingly less well for heavier atoms. By the time you got to a simple two-electron atom like helium, all of the clever rules from the Bohr-Sommerfeld model fell apart and did not help you accurately predict where the spectral lines should appear. Even the revolutionary discovery of matter waves seemed to be a technical curiosity that only seemed to shore up part of the foundation of the hydrogenic Bohr-Sommerfeld model. By 1925, the Bohr-Sommerfeld model was looking increasingly incomplete as new features in atomic spectra started to crop up that had not been properly explained.

A wave theory for quantum physics

Between 1925 and 1926, two major approaches to atomic physics appeared that each went a long way to patching up the problems with the so-called Old Quantum Theory of the atom offered by Bohr and Sommerfeld. These two ideas were called Wave Mechanics and Matrix Mechanics, and seemed to have sprung up overnight from the minds of two creative theoretical physicists. Each gave equivalent descriptions of hydrogen atom structure but approached the problem from very different perspectives. Their combination and refinement is what we now call the New Quantum Theory or simply 'quantum mechanics'.

The Austrian physicist Erwin Schrödinger took de Broglie's ideas to heart and developed an entire theory of the atom based on the behaviour of waves. The technical details of how he did this are the typical subject of study for university physics students because it is, as you can imagine, highly mathematical. It also introduces a whole host of new ideas to make the calculations of what quantum matter was doing easier to perform. However, it is also an approach that grew to completely dominate how modern physicists make predictions about what to expect in the quantum world. The Bohr model did not provide a detailed mathematical description for where the electron was located in the atom. All Bohr needed was for the electron to be in a specific orbit in the atom, but the details of exactly what it was doing in that orbit were irrelevant. This omission was remedied by Schrödinger because the de Broglie wave could not be confined to a mathematically precise orbit but had to be spread out as a wave in some way. But how? Schrödinger did what any mathematician would do. He represented the electron by what is called a mathematical function whose form was to be determined by solving an equation.

ERWIN SCHRÖDINGER

Erwin Rudolf Josef Alexander Schrödinger (1887–1961) was born in Vienna, Austria, and studied physics in Vienna, receiving his Habilitation certificate in 1914. Following his service as a commissioned officer during World War I, in 1921 he was awarded a professorship at the University of Zürich.

His early experimental work on atmospheric radioactivity and light were soon eclipsed by his theoretical interest in the structure of the atom through his collaborations with Wolfgang Pauli and Arnold Sommerfeld. This led to a series of papers in the early 1920s where he analysed the orbits of electrons in terms of their geometric features. This led to his publication of the 'Schrödinger Equation' in January 1926.

In 1933 he decided to leave Germany because he disliked the Nazis' antisemitism, but following unsuccessful stints at Princeton University, and several offers from the University of Edinburgh and Allahabad University in India, he returned to Austria in 1935 and a position at the University of Graz. Because of his anti-Nazi beliefs, he lost his position at Graz in 1938 and fled to Italy, later taking positions at Oxford and Ghent University.

He and Paul Dirac were awarded the 1933 Nobel Prize in Physics for the discovery of 'new productive forms of atomic theory'. Following his work in theoretical physics, his later years found him studying diverse subjects including

the perception of colour, free will, the mind-body problem and the search for a unified field theory of physics. After many lifelong bouts of tuberculosis, he finally succumbed to the disease in 1961.

Erwin Schrödinger.

A mathematical function is a recipe for converting one number into a unique second number. For example, in the simple function $f(x) = 3x+1$, the value of the variable x is found by solving the function $f(3)$ so that if $x = 3$ then $f(x) = 10$. Functions can have more than one 'independent' variable like the x in the example. In 3D space, the three coordinates for a point expressed as (x,y,z) can lead to functions in three independent variables such as $f(x,y,z)$, and if you add time as a fourth variable you can get $f(x,y,z,t)$ and so on.

Schrödinger's function could not simply be guessed at but had to have a form exactly determined by solving an equation. Schrödinger's function is like the variable 'x' in the above equation but instead of returning just a single value as a solution, the equation returns a precise description for the formula (function) describing the way the electron wave is spread out in space and time. Think of the path of a tennis ball after it is struck being defined by one set of equations, and the solution to these equations is the mathematical function defining the parabolic path of the ball: $f(x,y,z,t)$.

Because the electron was a de Broglie wave, the function had to have the properties of a wave and the equation that needed solving was, of course, called a 'wave equation'. The wave equation would describe how the wave function changes in time and space for a given set of conditions. These conditions included the energy, n, and angular momentum, l, quantum numbers used by Bohr, but also included a new quantum number, m, added by Sommerfeld that took care of the observations of the Zeeman Effect. This effect appeared in spectroscopic studies of atoms when they were placed in a strong magnetic field. Instead of one line appearing as an electron jumped from one energy level (n,l) to another, two lines appeared. This doubling could be explained by adding a third quantum number, m, to define the state of the electron (n,l,m).

THE MOST BEAUTIFUL EQUATION IN PHYSICS

The Schrödinger Equation is sometimes called one of the most beautiful equations in physics; it is certainly one of the most mysterious! Let's try to de-mystify this equation by looking at each of its pieces.

$$-\frac{\hbar^2}{2m}\nabla^2\Psi + V\Psi = i\hbar\frac{\partial\Psi}{\partial t}$$

The Schrödinger Equation has a number of familiar elements to it. The wave function that I have been talking about appears as the majestic Greek symbol, Ψ. The goal of solving this equation is to find what the wave function is. This is kind of like solving 'x' in the secondary school equation $x^2 + 2x + 1 = 4$. The first term in the equation on the left is negative (-), and has to be added (+) to the second term in order for it to equal (=) the term on the right side of the equals sign. There is also a curious quantity shown as the letter h with a bar through it ($\hbar$). This is nothing more than Planck's constant divided by the quantity 2 ϖ, which numerically equals 1.05×10^{-34} joule-seconds. Whenever you see Planck's constant in any equation, you immediately know that you are dealing with an equation that is trying to describe something in the quantum world. The next symbol you see is the letter 'i', which is the mathematical short-hand for $\sqrt{-1}$. This is a hint that something about the way the wave function is defined requires imaginary numbers. The last two familiar symbols are 't', which represents time, and 'm' which represents the mass of the electron, which has the numerical value of 9.11×10^{-31} kilograms.

The form of the wave equation mimics the total energy equation used to describe ordinary bodies in which you add the kinetic energy of the particle to its potential energy to

get its total energy. The first term in Schrödinger's Equation
is the equivalent kinetic energy term for the electron wave.
The second term is the potential energy, V, of the electron
wave, and the right side of the equals sign is the resulting
total energy. If you know the shape of the wave function,
you can plug it into the equation and compute the electron's
equivalent kinetic and potential energies along with its
total energy. Alternatively, if you know the shape of the
potential energy function, V, and the electron's total energy
and momentum, you can calculate what its wave function
should look like for a particular orbit in the atom. This all
sounds a bit complicated, but the details can be described
graphically. The most important thing is what the wave
function represents, and this is where Schrödinger's wave
mechanics is so intuitive to physicists.

The physical nature of the wave function

The wave function, Ψ, contains all of the information you can
ever know about the electron under the conditions of where it is.
If the electron is inside a hydrogen atom, Ψ gives you all of the
possible orbits the electron can ever have, and in which it can
ever be observed. For example, we can take Sir Isaac Newton's
model for our solar system that predicts the orbits for any planet.
Because planets can be found anywhere in orbit around a star, his
general 'solution' for an eight-planet planetary system contains
an infinite number of possible examples. But our observations of
our own solar system immediately reduce this large number of
possible 'states' for eight planets to only the one state we see in
our solar system today. Schrödinger's wave function, Ψ, is very
much like that. It offers an infinite family of possible models for
what the electron is doing. Each of these possibilities is called
a quantum state and is represented by the specific values of the
three quantum numbers, n, l, and m. Given the integer values

for n, and l used in the Bohr-Sommerfeld model, and the new number, m, to handle the Zeeman Effect, the wave function Ψ can be simplified into a specific function that gives the value of the state at each position in space (x,y,z) and at a moment in time, t. This can be plotted and the resulting picture corresponds to the detailed shape of the Bohr orbit in three-dimensional space for that specific state.

The success of Schrödinger's wave function approach was that it experimentally matched the model by Bohr-Sommerfeld and predicted the placement of the hydrogen spectral lines exactly, even including the Zeeman splitting effect controlled by the quantum number: m. Schrödinger's approach and calculations published in 1926 went far beyond the Bohr-Sommerfeld approach because it let you actually calculate where the electron would likely be located inside the atom, and revealed by its use of de Broglie's matter waves that many different standing waves were possible and only a few looked even vaguely like the Bohr-Sommerfeld closed, elliptical orbits at all. The success of the Schrödinger wave function analysis also let physicists use this mathematics as a new microscope into atomic structure, but there was a catch to this success, and that had to do with exactly what the wave function, Ψ, was physically representing.

Schrödinger's idea was that Ψ was a quantity that tells how much electric charge from the electron was present at a particular location in space – called the charge density. In other words, the electric charge of the electron was not confined to a single particle, but was taking advantage of its de Broglie wave-like aspects to spread itself out in space. The problem with this interpretation is that, although charge is a real quantity, the wave function Ψ yields numbers that are imaginary and involve $i = \sqrt{-1}$. That means Ψ is not actually a measurable physical 'real' quantity, but it is possible to make it produce real measurable numbers. If you calculate $\Psi^*\Psi$, which is a mathematical operation you can do on any imaginary number, you make it a real number. For example, if $\Psi = 1+3i$, then its complex conjugate $\Psi^* = 1-3i$, the product

$\Psi^*\Psi = (1-3i)(1+3i) = 10$. So it isn't Ψ that represents the density of the electric charge in space but $\Psi^*\Psi$. Although this solved the problem of how to physically interpret Ψ in terms of a real, measurable property of electrons, there was a deeper problem. If $\Psi^*\Psi$ represented the charge of the electron distributed in space, the charge is negative everywhere and that would cause the cloud of points mapped by $\Psi^*\Psi$ to repel itself and explode the hydrogen atom. Almost at the same time Schrödinger published his mathematical methods and interpretation for Ψ, physicist Max Born published an interpretation of Ψ, that completely eliminated this explosive problem, and is the modern 'gold-standard' definition we use today.

According to Born, the product of Ψ with itself, $\Psi^*\Psi$, is indeed a real number, but it is related to the *probability* of finding an electron at a certain location in space and time. The quantity Ψ is called the 'probability amplitude' to distinguish it from the actual probability. When you add up $\Psi^*\Psi$ over all space you get a number that is precisely 1.0, which Schrödinger interpreted as 1-unit of electric charge. What Born proposed was that this 1.0 represented a probability of 100 per cent that the electron would

be found somewhere in space if you went looking for it. The Schrödinger equation lets you calculate the mathematical form of Ψ. To then determine where the electron is within the atom you calculate its cloud of probability given by $\Psi^*\Psi$. Wherever that cloud has the highest values, that is where the electron has the highest probability of being for a given quantum state.

Max Born.

This probability idea was entirely new to physics because, in the past, the only things you calculated were real things like the strength of gravity, a position in metres or the intensity of an electric or magnetic field. What Schrödinger's equation calculated was essentially how probability acted like an almost physical object that obeyed a wave equation like sound waves or electromagnetic waves, but probability is not a physical thing! Probability is a statement of whether something will happen or not, or whether you measure a particular physical quantity or not. It was never considered to be a thing-in-itself like the butter on a piece of bread. But physicists are enormously practical folks. If a theory works and makes accurate, testable predictions, some physicists will use it and not worry too deeply about what the basis for the calculation method really is. Other physicists were enormously troubled by this theoretical newcomer, the probability amplitude, and even today the proper interpretation of what the wave function Ψ means seems to elude a simple answer. One thing is for certain, however, and that is that the interpretation of the wave function is rooted very deeply in the simple act of determining where an electron is located and what its properties are; in other words, the very act of observation itself.

The wave function and quantum states

One interesting feature of Ψ is that it is a Trojan Horse. It isn't just one mathematical expression but is an infinite series of expressions:

$$\Psi = a_f(0,0,0) + a_f(0,0,1) + a_f(0,1,0) + \; etc.$$

Solving the Schrödinger equation specifies the general form for Ψ in terms of an infinite series of other functions, $f(n,l,m)$, that each represent one of the possible quantum states for the electron. If you want to determine what the probability is for observing the electron in the quantum state with $n = 0$, $l = 1$ and $m = 0$ it is represented by the third term in the series, and the probability will be calculated from the complex number a3 as $P = a3^*a3$. If you then want to map out what the electron's probability looks like in space, you just plot $f(0,1,0)^*f(0,1,0)$. But there is an interesting rub to this interpretation.

If you do *not* try to observe the electron inside the atom by, for instance, bouncing a photon off of it, the wave function Ψ gives you the exact information about where it is located and in what possible states. The wave function contains a *huge amount* of information about the possible quantum states for the electron in the atom. But at the instant you make the observation, all of this nearly-infinite amount of information vanishes and you are now left with much more limited information about the electron being in a specific quantum state such as $n = 2$, $l = 1$ and $m = 1$. In addition, now the electron is definitely known to be at one location in the atom given by its space coordinates (x,y,z).

Physicists are currently trying to understand two things about this process. First, where did all that information go after you made the observation? Also, exactly how did the physical electron change from a wave-like object treated by Schrödinger's wave mechanics to an object with a definite particle property at the point and instant of measurement? The devil is in the details, as they say, and these details are among the most vexing issues in modern quantum theory today, which I will discuss in the next chapter on Heisenberg's theory. Meanwhile, Schrödinger's wave theory had the additional benefit that, not only did it allow for the first time the calculation of electron quantum states inside atoms and what physical shape they would take, but it also explained the mechanism of radioactivity. This phenomenon occurs because sometimes a quantum particle (alpha particles for example) can be found outside the system (atomic nucleus) that we think is containing it.

You can think of an atomic nucleus as a closed box. The nuclear particles are trapped inside because, like miniature rockets trying to leave a planet, they do not have enough energy to break free from the nuclear forces. That is how we could think of the problem thanks to 300 years of learning physics from Sir Isaac Newton. But quantum particles are not like solid objects. They also have a wave-like character. Seen in terms of Schrödinger's equation, there are three different regions in space defined by the nucleus. There is the interior where the nuclear particles live as trapped particles, there is an outside zone where particles are free to move unbounded by the nuclear forces. Then there is a third 'barrier' region in-between the two near the surface of the nucleus where a particle would be able to transition from a bound to a free particle and where it is neither bound nor free. Because we are dealing with matter as a wave, the wave function, Ψ, has to smoothly change its shape if you start from the bound region and trace it into the free region outside the nucleus. What does this all have to do with radioactivity? No matter how you set up Schrödinger's equation to represent the nucleus of the atom by adjusting the 'V' in the equation, the wave function will start out large inside the nuclear zone, diminish to a smaller value as it passes across the barrier zone, and then maintain a small but non-zero value as it enters the free zone. This means that the probability $\Psi^*\Psi$ is large that the particle is still trapped inside the nuclear zone, but it is absolutely not exactly zero in the free zone. There is some probability that the quantum particle in the nucleus has actually leaked out of this confinement and escaped the nucleus entirely. There is no way to duplicate this sneaky behaviour using any version of Newton's physics. It is a completely quantum mechanical property of matter and is called tunnelling.

Examples of quantum tunnelling

The idea that particles can escape bound systems even when they do not have quite enough energy as predicted from Newtonian physics was a huge surprise to physicists. In the modern era, our smartphones and other advanced devices would never have been possible without this phenomenon. Inside every smartphone are electronic components called MOSFET transistors that operate by controlling the tunnelling current passing through the physical barriers that form their junctions. Other devices called tunnel diodes, which you can purchase for a few pounds, are used in many applications involving microwave communication. But perhaps one of the most 'glaring' applications is to be found in the sun itself.

QUANTUM TUNNELLING

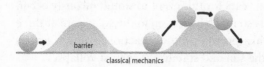

classical mechanics

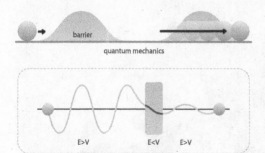

quantum mechanics

The wave function inside an atomic nucleus has a large amplitude (probability), but is not exactly zero outside the energy barrier region between x = 0 and X = L.

For generations, scientists considered different schemes whereby the sun would create its light and heat. In the early 1920s Sir Arthur Eddington proposed that the source of the sun's energy to stabilize it against gravitational collapse was the fusion of hydrogen into helium. This idea was generally rejected because the centre of the sun at 15 million Celsius was too cool for individual protons to collide and fuse against the tremendous force of electrostatic repulsion between like charges. With the advent of quantum theory, Hans Bethe in 1939 applied the idea of matter waves to the proton wave functions during a collision event and discovered something remarkable. Even though they may not have had enough kinetic energy to overcome their mutual repulsion, there was a very small probability that at the temperatures of the solar interior protons could in fact fuse together to form a deuterium nucleus. This would not happen with every collision, however, because one of the protons would have to simultaneously turn into a neutron to make a stable deuterium nucleus. This circumstance requires what later became known as the weak interaction so that a given proton would have to wait as much as 10 billion years for this event to simultaneously occur. Nevertheless, there is so much hydrogen in the solar core at these temperatures that this 'proton-proton' reaction is just vigorous enough to light up the sun and stabilize it against collapse.

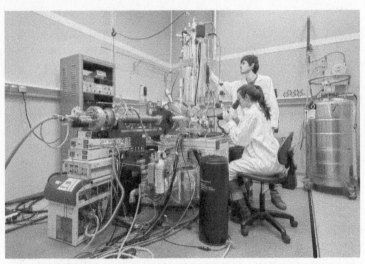

A scanning tunnelling microscope (STM) takes advantage of the tunnelling current that flows between an object and the STM's sensing needle to form images of individual atoms.

CHAPTER 7:

Heisenberg's Uncertainty Principle

In no other subject than philosophy are we ever confronted with the kinds of issues raised by quantum mechanics, especially when it comes to the simple act of observation. For sighted people, this can be summed up by the question 'Would the moon exist if there was no one to see it?' For the sight-impaired person, it is the question 'If a tree falls in the forest and there is no one there to hear it, does it make a sound?' Neither of these koans are to be taken literally, but instead are asking the deep question about whether the existence of a perceivable thing depends completely on its ability to be perceived. This is a deeper problem than what you might suspect. Consider, for example, the entire idea of a past history. There is no observer present today who observed and 'made real' by this act any single historical event from more than 140 years ago. This zone extends all the way to the emergence of dinosaurs, the formation of our planet earth, and the Big Bang itself. Even the events of the present moment in which you exist could be corrupted by this demand that the only real things that exist are things that are observed. Werner Heisenberg took the extreme position that objects do not exist except at the instant of being observed. So to ask what an electron is doing when it is not being observed is a question without any meaning. Erwin Schrödinger took the more measured view that identified the wave function as being real, but its value 'collapses' at the moment of measurement into a specific state among an infinite number of possibilities.

The nature of the measurement process

The solution to this existential problem was that God was made the eternal and omnipresent observer who made sure that things existed even when lowly humans were not even around to observe them. He can peer into untrodden dark corners of history where humans have not yet been and make phenomena there come into existence. But the question still remains in modern physics today in the post-quantum world: are quantum particles, even macroscopic particles, real even if we do not measure them? What exactly happens when we make a measurement? If you measure some property and do not tell anyone about the result, do you actually now live in a completely different reality than other people because they don't share with you this bit of information?

With Schrödinger's wave theory, an electron in an atom could be in any of an infinite number of quantum states seemingly all at once, at least that's what the mathematics says. Once an electron was measured (observed) by some instrument, all of these possible quantum states disappear like ghosts, leaving behind only one state. But, had you not made that measurement, it would still be in its own reality as a mixture of millions of different quantum states all at once. Your reality as its observer started out as a superposition of millions of different possible states for the electron, but when you made that fateful measurement, all of your possible measurement outcomes collapsed into just one state. This phenomenon, called the collapse of the wave function, was a very vexing issue implied by Schrödinger's mathematics, but the success of his theory in computing numerous experimental outcomes was too compelling to simply dismiss because of some abstract philosophical concern. As it turns out, another completely different approach had been taken almost at exactly the same time by a very young German physicist by the name of Werner Heisenberg.

WERNER HEISENBERG

Werner Karl Heisenberg (1901–76) spent much of his late-teenage years reading classical literature and hiking in the Bavarian Alps, which is where his formative ideas would lead him to his later realizations that 'Modern physics has definitely decided in favour of Plato. In fact the smallest units of matter are not physical objects in the ordinary sense; they are forms, ideas which can be expressed unambiguously only in mathematical language.'

He studied physics and mathematics at the Ludwig Maximilian University of Munich where he studied with Arnold Sommerfeld and later at the University of Göttingen where he studied under Max Born and David Hilbert and received his Doctorate in 1923 working on problems in turbulence. In 1924 he travelled to Copenhagen to work with Niels Bohr on the structure of the atom. During a six-month period in 1925 he collaborated with Max Born and Pascual Jordan at Göttingen on an approach to atomic physics he called matrix mechanics.

This led in 1927 to his development of the idea of the Uncertainty Principle which now bears his name. In contrast to Einstein and de Broglie, who believed that particles always had a well-defined position and momentum, Heisenberg as

an anti-realist believed that, like Plato, particles are not real in the usual sense but are only constructs with ill-defined properties. Their 'reality' if it exists at all is beyond the scope of science and direct observation to determine.

Werner Heisenberg.

Heisenberg's philosophy about the quantum world was actually very simple. Only the things that you can measure are real. Everything else is *irrelevant*. If an electron travels from Point A to Point B, it is only what measurable effects it makes at Point A and Point B that should be the subject of physics. Everything else is a fiction. In real, measurable terms, what the electron was doing between Point A and Point B was not measurable and knowable, so in a real sense the electron can be said not to exist between these points so far as Heisenberg was concerned. Heisenberg's obsession with only the properties of phenomena that were measurable was liberating because you didn't have to worry about how an electron was moving inside an atom. Its path was irrelevant because it was not one of those things you could measure. Only the outcome of specific measurements, such as the wavelength and intensity of a specific spectral line, were relevant to physics. You could measure an electron's three quantum numbers for energy, angular momentum and its magnetic quantum number, but these three numbers, themselves, were the only things you could measure. How or where the electron was inside the atom just before or just after the measurement was made was not observable and so in a sense the electron did not really exist in some state of motion inside the atom. Whatever it was doing to manifest the three observable quantum numbers was not really of any interest.

Heisenberg's challenge was to come up with a mathematical theory that described this state of affairs so that you could actually launch predictions from it and confirm these ideas. The approach he took was to think of an atom as a vast table of possible observations. Each cell in the table gave a probability for a particular combination of quantum numbers to exist at the time of a specific measurement, and it was these tables, called matrices, that gave this theory the name Matrix Mechanics. In a later conversation with Wolfgang Pauli he said 'Everything is still vague and unclear to me, but it seems as if the electrons will no more move on orbits.'

What Heisenberg had been struggling with had to do with calculating the intensities of the spectral lines in the hydrogen atom. Bohr and Sommerfeld had shown how to calculate their wavelengths, but there was nothing that hinted at how bright these lines should be. Heisenberg's ideas went to a place that scientists do not like to go. It also reinforced the idea so common among non-scientists that scientists only believe in things that can be seen and measured. Let's see just how radical this idea is by a simple analogy that Heisenberg liked to use.

'It was about three o' clock at night when the final result of the calculation lay before me. At first I was deeply shaken. I was so excited that I could not think of sleep. So I left the house and awaited the sunrise on the top of a rock.'
Werner Heisenberg, May 1925.

Heisenberg's view of reality

Imagine yourself sitting on a park bench at night. There are the occasional lamps here and there lighting the path. Suddenly you see a person standing under one of the lamp posts. The person seemed to appear out of nowhere, then after a few moments the person disappears into the darkness and reappears at another lamp post a few hundred metres away. You don't know exactly what path she took to get from one lamp post to the other because there are many paths that could have been taken, either across the lawns or along the cobblestone path. What you do know is that when she arrived at the first lamp post you made a set of observations and determined that she looked at her watch. At the second lamp post you made another set of observations and determined that she opened her purse and took out her mobile phone to make a call. Heisenberg believed that all we can determine about the existence of an electron was what it was doing at the 'lamp posts' where an interaction, observation or measurement occurred. How it arrived at each of these locations was hidden beneath the darkness of impossible measurements that could never be made. In the hydrogen atom, all of the possible measurements could be tabulated to give a probability or 'intensity' of the measurement if it were to occur. But there was no need to worry about what an unobservable electron was doing in between the measurements. For all intents and purposes, the electron only existed at the instants, locations and quantum states when it was measured. Well, this is a rather evocative description of electron physics, but how could you possibly convert this picturesque idea into something that could be mathematically computed and give a concrete prediction that was testable?

When a hydrogen atom emits or absorbs a photon, the electron in one stable quantum state suddenly jumps to another quantum state. Heisenberg was not interested in what the electron was doing as it was changing between the two quantum states because, like the lamp post analogy, the electron did not exist during this in-between time. It simply absorbed a photon in one state, and 'jumped' into the next quantum state. So for Heisenberg, what was important was calculating what were called the transition probabilities in going from state A to state B. But not only did he have to calculate these in terms of the wavelength of the photon, which was a measurable property, but he had to derive the intensity of the light, which was also a measurable property. When you look at the spectrum of hydrogen, it is only the wavelength and intensity of the spectral line that are the 'observables' for the quantum system involving the electron inside the hydrogen atom.

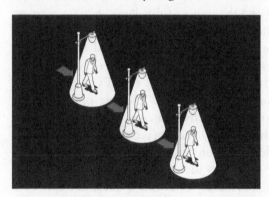

Heisenberg proposed that electrons only exist 'under the lamp posts' and not in the dark spaces in-between.

Heisenberg had been working on other problems with his mentor, the Dutch physicist Hans Kramers, who had worked with Niels Bohr in Copenhagen to get his doctor's degree in 1919. What Kramers had done was to mathematically analyse the precise electron orbits in the Bohr-Sommerfeld model by a method called Fourier analysis. This mathematical technique replaced each orbit by an infinite series of wave-like terms that were added together to recover the original, precise orbit. But along the way, the amount of each term that you had to add in order to recover the circular path was related to the 'intensity' of the orbit, so when an electron jumped from one orbit to another, it was hoped that the difference in these Fourier term intensities would be related to the intensity of the resulting spectral line.

Unfortunately, the predicted wavelengths for these orbit terms did not match up with the observed spectral line wavelengths. But in a moment of insight, Heisenberg realized that Kramers' approach might actually lead to the real wavelengths and intensities if he replaced Kramers' Fourier terms with a table of numbers instead of one specific value as the Fourier method required. By doing this, he could smear-out the information in each state to make the state look more like a quantum state, and hopefully give a prediction for the intensity of the spectral lines produced by each jump. The technical details for how Heisenberg worked out all these details is beyond the scope of this book, but I can try to simplify some of the important steps.

Through some trial and error, Heisenberg discovered that these new numbers, called matrix elements, were actually pairs of numbers, one for electron position, called Q, and one for electron momentum, called P. It was the size of the matrix elements for P that were directly related to the intensity of the spectral line. Each matrix element gave the probability that the electron would transition from an initial state to a final one and this was related to the intensity of the spectral line. Heisenberg's matrix elements for P and Q are also dependent on the electron's energy level so that the rows of the matrix, n = 1,2,3 etc, represent the initial state energy and the columns, m = 1,2,3 etc, represent the final state energy. We can write any

Hans Kramers (centre) with fellow physicists George Uhlenbeck and Samuel Goudsmith.

of these matrix elements using the short-hand symbols Pm,n and Qm,n. For example, if you wanted to know the position of the electron for a particular final state, say m = 3, you had to sum up all the matrix elements in the Q matrix along the n-row with that value for m so that Q3 = Q3, 1+Q3, 2+Q3, 3+etc. To calculate the measured intensity of the spectral line that had the energy difference of hv = Em-En, all you had to do was calculate the matrix element of Qm,n.

The Heisenberg Principle of Uncertainty

Heisenberg's matrix mechanics was very cumbersome to work with, mainly because physicists at that time had never worked with the mathematical concept of matrices. His ideas were not only inherently unfamiliar, that electrons only existed when they were measured, but his mathematical tools were also inaccessible to many physicists. Since Schrödinger's wave mechanics was far more intuitive and easier to mathematically work with, the Schrödinger theory won out. Heisenberg's deliberations were not, however, completely in vain. Because Heisenberg's ideas were rooted in only observable features of the quantum world, this led to tremendous new insights about what the act of measurement entailed, along with its limitations.

The challenge that Heisenberg had encountered was that both the P and Q matrices had to obey a specific multiplication rule. If you took matrix P and multiplied in by matrix Q, that did not give the same answer (also a matrix) as when you multiplied matrix Q by matrix P. He found from his quantum considerations a very peculiar rule that QP-PQ = iℏ. What this simple relationship means is that there can be no quantum states where the electron simultaneously has a definite measurable position and definite measurable momentum. In fact, when written with position and momentum being real numbers and not matrices, you get

$$\Delta q\, \Delta p \geq \frac{\hbar}{2}$$

This inequality has the interesting property that, if you made a definite, precise measurement of the position of a particle, that means the uncertainty in your measurement is just $\Delta q = 0$. But if we solve the equation $\Delta p \geq \hbar/(2\Delta q)$, if $\Delta q = 0$, then Δp must be infinite. In other words, your precise measurement of the particle's position, q, forces you to give up any precise knowledge of the particle's momentum, p, at the same time. You can know where the particle is, but in the very next instant you cannot know where the particle is going. This is called the Heisenberg Uncertainty Principle (HUP), and it is a foundational principle in all of modern quantum mechanics. Amazingly enough, you do not need only to consider quantum particles to see how this works.

In an earlier chapter we saw how Thomas Young's single-slit experiment demonstrated the diffraction and interference of light. In the single-slit experiment, a light ray is sent through a narrow slit in a barrier, and then strikes a screen on the other side to reveal a diffraction pattern. This pattern has a main, intense line parallel to the slit, followed by weaker lines to either side. As you make the slit narrower and narrower, the width of the lines increases. What is happening is that as you narrow the slit, you are decreasing the uncertainty Δx in the position of the photons as they travel through the slit. But the pattern on the screen is determined by the momentum of the photon perpendicular to the slit axis. Because the diffraction pattern is spreading out as you decrease the slit width, this is saying that as you make the position measurement of the photon more precise, your knowledge of the momentum Δp of the photon is getting worse. The reason this happens can be illustrated by using the mathematical technique of Fourier analysis (see box).

What this Fourier decomposition means for light and for quantum particles is that, if you want to know exactly where the particle is by narrowing the slot, you will have to add more terms to the Fourier series, which means adding more light waves with differing frequencies. But because wave frequency is related by Planck's Law to the energy carried by the photon, you are losing information of the definite energy/momentum carried by the photon (making ΔP larger) and at the same time you are increasing your knowledge of exactly where it is (making Δx smaller). This is the same as stating Heisenberg's Uncertainty Principle.

FOURIER ANALYSIS

It's a daunting name but the basic idea is pretty easy to show, graphically. Back in 1807 the French mathematician and physicist Jean-Baptiste Joseph Fourier was exploring trigonometric functions like the sine and cosines we learned back in secondary school. What he discovered is that any mathematical function that was smooth, could be represented by adding together a series of sine or cosine waves each with the right heights (called amplitudes) and wavelengths. One important example is the slot function, sometimes called an impulse, which has a value of zero everywhere except between the two walls of the slot. The Young experiment uses a slot which only permits light to pass through a barrier. Another example of a slot is the binary code used by computers, which has a value of zero except at a time interval when the binary bit is set to a value of 1. Because radio signals and computer networks use electromagnetic waves to transfer information (voices, music and data bits), electrical circuits have to be designed to detect on-off information at very high speeds. If we take a slot and perform a Fourier analysis on it (called a decomposition) we get a series of sine functions. The more of these we add together, the more closely the sum resembles the slot where it is zero everywhere except between a specified range of positions or times.

$y = sin(t) + sin(3*t)/3 + sin(5*t)/5 + sin(7*t)/7 + sin(9*t)/9 + etc.$

The act of observation and uncertainty

The way that HUP influences the very act of measurement was also a significantly new discovery. We learn about where things are by looking at them. This act involves looking at the light reflected by them. In our new language of photons, we shoot a photon at an electron and from the photon that is reflected back to us we determine that the electron is located exactly at a specific location in space. To make this position measurement, we have to interact with the electron by striking it with a photon. Now, the photon can have any wavelength we choose, but the accuracy of our position measurement will be limited by the photon's wavelength. If it is 1,000 nm, then we will not know where the electron is to better than 1,000 nm. So, we use a photon with a wavelength of 1 nm to get an even more precise position measurement. We know from Planck's Law that the energy of a light quantum depends on its frequency according to E=hv, but because for light $v\lambda=c$, that means $E = hc/\lambda$. As we decrease the wavelength of the photon to make ever-more precise position measurements, we have to strike it with photons of increasingly higher energy. This collision has the consequence that as we gain better knowledge of position (making Δx smaller), we lose better knowledge of where the electron goes next because we can never tell exactly what the outcome of the collision is with enough precision to make the uncertainty in the momentum (making Δp larger) of the electron get smaller at the same time.

$$\Delta x \, \Delta p \geq \frac{\hbar}{2}$$

What HUP does is to place a limit on how well we can know both position and momentum at the same time in any interaction, especially those involving our attempt at measuring position and momentum with the highest accuracies. There is also a similar relationship between energy ΔE and time Δt that places a limit on how well you can know how much energy a particle has over a specific time interval. The shorter you try to make this time interval to get the best possible measure for the particle's energy, the less precise will be your estimate for the energy.

$$\Delta E \, \Delta t \geq \frac{\hbar}{2}$$

The HUP applied to a particle's energy has an even more peculiar consequence in quantum mechanics. Suppose you create an 'empty' box with no particles in it. Because according to Einstein's theory of relativity, $E = mc^2$, you have removed all of the particles from inside the box that carry mass, m, and reduced the energy of the box's contents to exactly $E = 0$. But the precision of this energy measurement is set by HUP, so that you only know that E = 0 with a specific uncertainty ΔE. The HUP energy relationship re-arranged with a little algebra says that $\Delta E \geq \hbar/2\Delta t$, which means that in order to make the uncertainty in the energy measurement ΔE as small as possible, you simultaneously have to increase your uncertainty in knowing exactly over what time interval Δt your box actually contained no particles. I am going to come back to this energy-time relationship in a later chapter because it provides an explanation for many other things in the quantum world and is essential for understanding them.

CHAPTER 8:

The Hydrogen Atom Reborn

Schrödinger's wave function has a lot in common with the phenomenon of standing waves in nature, except that they do not represent a physical medium such as sound, air or water. Instead of being measured by properties of a medium such as density, height or speed, which are familiar physical properties, they are measured in terms of the square-root of probability. Physical standing waves are measured by instruments that sense density, height or speed. For a given standing wave only a small number of measurements need to be made to define the shape of the wave. Wave functions, however, are related to whether an observation will detect the particle at a specific point in space and time, with a specific energy and momentum. To map out the wave function of a single hydrogen atom you have to make individual measurements on thousands of atoms containing electrons in exactly the same quantum state to define the probability of detecting an electron at specific locations.

The most useful atom in the universe

Hydrogen is the workhorse of chemistry on this planet, for the simplest of all reasons: it is the most abundant element in the entire universe. Cosmologically, there are three hydrogen atoms for every helium atom, but the remaining 100+ elements known today are mere impurities accounting for far less that the remaining 1 per cent of all matter in the universe. Ironically, although hydrogen and helium are so abundant, it took thousands of years before we even knew they existed.

In the early 1500s alchemist Paracelsus noticed that when iron filings were added to a solution of sulphuric acid, it would give off a flammable gas. Although Robert Boyle produced hydrogen gas in the experiments he conducted in 1671, he didn't follow up on this result. It would take nearly another 100 years before Henry

Cavendish announced in 1766 that it was its own unique element. Its name, hydrogen, was a contraction of hydro-gene or 'water former' and was coined in 1783 by the French chemist Antoine Lavoisier.

Antoine Lavoisier.

Helium, meanwhile, had its own interesting history because it was actually discovered on the sun not on earth. In 1868, the French astronomer Pierre Janssen travelled to India to observe the solar eclipse using the new technology of the spectroscope. Among the numerous Fraunhofer lines he found a new yellow line, which he noted as possibly a new element. The English astronomer Norman Lockyer was also familiar with using spectroscopy to study astronomical objects. Viewing the same eclipse, he independently spotted Janssen's yellow line, proposing the name 'helium' for it. It would take another 20 years but eventually signs of this element turned up in the gases emitted during an eruption of Mount Vesuvius in 1882, discovered by the Italian physicist Luigi Palmieri.

THE POWER OF HYDROGEN
Hydrogen and its sister element helium inevitably found their way into commercial and industrial applications. No sooner had hydrogen been named by Lavoisier in 1783 but on 2 March 1784 Jean-Pierre Blanchard made

The *Hindenburg* disaster of 1937.

the first manned balloon flight using this lighter-than-air gas. It would be another 120 years before Count Ferdinand von Zeppelin launched the first hydrogen-filled commercial air ship in 1900 called the Zeppelin LZ5. The succession of long distance and trans-Atlantic flights by Zeppelins, or 'dirigibles' as they were known in England, ended spectacularly with the deadly crash of the LZ-129 Hindenburg in 1937. The usefulness of this gas increased considerably from 1943 when it was discovered to be a potent rocket fuel. Hydrogen also became a critical element of gaseous fuel cells developed in 1842 by William Robert Grove, and later perfected for use in space exploration and transportation applications in the 1960s.

The hydrogen atom and standing waves

Meanwhile, studies of the spectral lines emitted by hydrogen turned out to be rich in clues about the structure of the hydrogen atom, now interpreted as a dense positive core called the proton, surrounded by a single electron that was responsible for the chemical properties of this element. At first, a planetary model was considered using the Bohr-Rutherford model, but the discovery of de Broglie matter waves led to a different interpretation. Heisenberg offered the revolutionary idea that it is meaningless to ask what the electron is doing inside the atom because there was no possible observation or measurement that could provide any insight to this question. But electrons could be liberated as cathode rays, so under other non-atomic conditions they were clearly observable and measurable. The Heisenberg interpretation was intriguing and revolutionary but did not lead physicists to concrete predictions, especially not ones that were easy to come by given the complexity of Heisenberg's mathematical 'matrix' techniques. At the same time, Erwin Schrödinger had created a complete and far simpler mathematical theory for how to describe the atomic electron and make quantifiable predictions. However, his approach forced physicists to consider the concept of the quantum wave function and its mysterious collapse to a definite state after an observation was made. What did these electron quantum states actually look like, and was there any way one could actually confirm that they existed in the way Schrödinger's waves required? Ultimately, these are mathematical questions requiring mathematical answers, but there are some examples from familiar nature we can use as a guide.

CHLADNI FIGURES

Ernst Chladni was a German physicist and musician interested in sound vibrations; in particular, ways in which you could make them visible to the eye. He borrowed an idea first proposed by Robert Hooke in 1680 to study vibrating patterns on a glass plate, and in 1787 published an entire book on his experiments: Entdeckungen über die Theorie des Klanges (Discoveries in the Theory of Sound). *His innovation over Hooke's was to sprinkle sand lightly over a metallic surface, and then run the string of a bow over the edge of the plate. The sound waves reverberating through the thin metal surface would instantly move the sand particles into complex patterns as the sand grains moved to places where the surface vibrations were minimal. The exact shapes of these figures are related to the shape of the plate (circular, triangular, square etc) and the precise location where the bowstring is struck, along with the speed of the string rubbing across the edge of the plate. Modern researchers and artists have discovered a variety of ways to excite these patterns into both beautiful and exquisitely complex shapes.*

When the pipe on an organ resonates to produce the low-C note on a pipe 2.5 m long, the air inside the pipe participates in a pressure standing wave that is one-dimensional along the axis of the pipe. This phenomenon is enough to create the rich experiences of hundreds of pipes 'stopped' to different lengths to produce all of the useful chromatic notes and their various octaves. There are also two-dimensional standing waves that are a bit less familiar but even more intriguing called Chladni Figures.

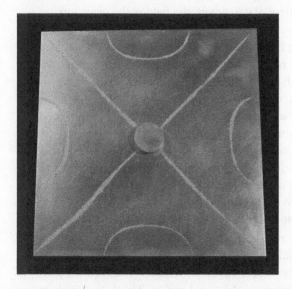

A simple Chladni figure caused by motions in a plate of copper.

Another situation in which two-dimensional standing waves can be found is on the surface of the sun. The interior of the sun is a thermonuclear furnace with heated plasma that is mostly quiescent until it gets close to the surface, where it starts to convect. This is a complex process that causes pieces of the surface to move outwards and inwards. Because there is nowhere for this plasma to go other than 'up and down' it produces a geometric pattern on the surface that can be studied to learn about the deep interior. These patterns are standing waves similar to the P-waves studied on earth during earthquakes, so this branch of astronomy is called helioseismology.

Examples of three-dimensional standing waves are hard to find when it comes to manipulating conventional media like water or air, which have acoustic properties. Instead if we think of light being a medium, there is one very common item you can find in your kitchen that has 3D standing waves: your microwave oven. Inside, electrons in the klystron tube cavity oscillate to produce an intense electromagnetic wave pattern inside the shielded oven box. Each of these places where the field is the most intense is a roughly spherical region about 10 cm in diameter. To avoid your food not being heated uniformly because part of it might be in the dark nodal regions, food is rotated on a turntable to keep it moving through the microwave peak regions. With a very simple experiment you can even 'see' where these locations are using a diode array. The diodes will light up if the microwave energy is high, and remain black if the energy is at a minimum.

The only other common example of three-dimensional standing waves are the quantum states within an atom. Recall that these are based upon de Broglie's idea of matter waves, and each quantum state is a specific pattern in 3D space that tells where an electron is likely to be. The matter wave follows the basic rules of constructive and destructive interference to create a stationary wave. Places called nodes where this quantum standing wave is absent are places where the electron is not likely to be, given its particular quantum numbers, n, l and m. The difference between quantum standing waves and the previous examples for acoustic, seismic and microwaves is that the quantum standing waves are not related to the intensity of a physical medium like electric field strength or air and rock compression and density. Quantum waves tell you about the probability of finding something at a particular location. We can use the wave function equation for a given state to create a mathematical model of where the probability for finding an electron in that state is highest or lowest inside the volume of the atom.

The observation of electron wave functions

Of course, with predictions from Schrödinger's theory it is a fair question to ask whether this approach to describing the hydrogen atom is more correct than other possibilities. The idea that the quantum states calculated by Schrödinger's wave mechanics have unique three-dimensional shapes and no others is something that has to be confirmed because it is just plausible that there are other possibilities offered by a 'better' theory than Schrödinger's. The challenge is that these wave functions are not physical things in space. They represent the combined outcomes of thousands or millions of possible measurements for where an electron in a particular quantum state could be located in space and time.

In 2009, physicist Igor Mikhailovskij and his colleagues at the Kharkov Institute for Physics and Technology in Ukraine succeeded in creating an image of a carbon atom revealing its lowest-energy probability clouds. The historic result was published in the prestigious journal *Physics Reviews B* on 7 October 2009. What they did was to use a new version of the field-emission electron microscope developed decades before, but with a sensing needle that consisted of a carbon nanotube held above a phosphor screen with a voltage difference of about 425 volts. The carbon atom at the tip of the chain emitted electrons onto the phosphor screen, which allowed them to form an image of the electron cloud around the carbon nucleus at the end of the nanotube carbon chain. It is important to understand that this image is the result of thousands of interactions between the electrons in the carbon atom and the phosphor screen. And so is a measure of the probability distribution cloud $\Psi^*\Psi$.

In the early 1980s, researchers at the Vinnitsa Polytechnical Institute in the former USSR developed a new kind of atomic microscope that involved carefully ionizing an atom and following the single emitted electron travelling at slow speeds onto the surface of a detector. Because the de Broglie wavelength of the electron increases as the electron slows down, working with slow electrons ensured that their matter wavelength was as large as possible. This also ensured that matter wave interference could be observed with the instruments. Many different paths could be taken by the electron and these different paths would cause an interference pattern on the surface of the detector. What was exciting about this new 'photodetachment microscope' is that the process did not upset the wave function of the electron. The resulting image from numerous separate trials would still contain preserved information about the original wave function of the electron when it was inside the atom. Many different applications of this technique were used during the intervening years until the early 2000s when a modification to this technology called the photoionization microscope was developed. The addition of an electrostatic lens allowed for greater spread in the paths of the electrons to the detector, essentially serving as a magnifying lens. At first, xenon atoms were used in 2001, but the recorded interference patterns were complex because of the many electrons involved within the xenon atom, so simpler atoms such as lithium and then hydrogen were finally used. In each case, what was being measured was the manner in which the magnitude of the probability density $\Psi^*\Psi$ varied with increasing distance from the atomic nucleus, called the radial distribution, which was known in detail from the Schrödinger equation. If the electron's wave function was undisturbed by the act of striking it with a carefully-pulsed laser beam, the pattern of electron interference at the detector should exactly match the radial distribution based upon the wave function prediction. In a series of two back-to-back papers published in the journal *Physics Review Letters* in 2013, a team led by S. Cohen at the University of Ioannina in Greece succeeded in creating this interference image for lithium atoms, and a second team led by Aneta Stodolna at the FOM Institute for Atomic and Molecular Physics in Amsterdam, used their photoionization quantum microscope to perform this feat for hydrogen atoms.

The image of the accumulated electron trajectories displayed the same pattern of nodes as expected for the radial distribution of the electron wave function. For the first time it was possible to verify the shape of the electron wave function, actually the probability density $\Psi^*\Psi$, within the atom for a carefully prepared quantum state, called a Rydberg State in which the electron is farthest from the nucleus but not quite ionized. This confirmation was heralded in many technical and popular articles around the world as a confirmation that quantum waves exist and are shaped in the way predicted by Schrödinger's theory. Although you can't image the location of a single electron inside the atom, through multiple measurements across thousands of atoms you can build up a probability map of where the electron should be in a specific quantum state, and this map matches the theoretical calculation in which the 'square-root' of this probability distribution is the actual wave function, Ψ. As experimenters continued to hone in on confirming the atomic probability distributions predicted by Schrödinger's wave mechanics, an entirely different set of investigations were tackling the related issue of what Bohr called the quantum jump.

The physics of a quantum jump

With Schrödinger's wave mechanics, when an electron changes from one state to another emitting a photon, the initial and final states are legitimate solutions to Schrödinger's equation that should persist indefinitely as stable states, so how does an electron make the switch from one state to another? Other than to say it happens instantaneously, Bohr's theory did not provide these details, and to make matters worse, Heisenberg's ideas about electrons implied that the electron leaves one state and arrives at the second state, but in-between does not exist at all.

In 2014, physicist Stephen Leone at U.C. Berkeley with collaborators in Germany and Japan used one laser to deliver visible light energy to silicon atoms in a crystal, then immediately used an x-ray laser with ultrashort pulses to take a series of 10 attosecond (1 attosecond is 10^{-18} seconds) 'snapshots' of what happened afterwards. The movie-like series of x-ray images was interpreted by a supercomputer model of this interaction to uncover the details. For the first 450 attoseconds, the electrons in the silicon atom jumped out of the atoms and entered the conduction region of the crystal, which then provides the carriers for the current in the semiconductor material. About 60 attoseconds after this jump, the crystal lattice begins to vibrate from the recoil of the silicon atom after the electron emission. So to make a quantum jump takes electrons 450 attoseconds (0.00000000000000045 seconds). For an electron with an energy of about 1 electron volt (eV) its de Broglie wavelength is 1.2 nm with a typical speed of about 2,000 km/s. The time for one de Broglie wave is about 500 attoseconds. So the time it takes for a jump is about equal to one oscillation of its matter wave. This was an interesting result that showed the jump time is regulated by the de Broglie frequency of the electron, but it did not give any details on what the electron was 'doing' within this 450 attosecond jump time.

ANETA STODOLNA

Aneta Sylwia Stodolna (b. 1986) studied at the Gdansk University of Technology in the Faculty of Applied Physics and Mathematics in Poland, receiving her Masters Degree cum laude in Applied Physics in 2008 where she investigated neutron-rich nuclei near the element lead.

Between July 2008 and March 2009 she worked at Katholieke Universiteit Leuven, Institute for Nuclear and Radiation Physics, Belgium, investigating the structure of exotic nuclei employing laser spectroscopy methods at the ISOLDE hall at CERN, Switzerland. In March 2009 she started her PhD research at FOM Institute for Atomic and Molecular Physics (AMOLF), Amsterdam, the Netherlands, in the Extreme-Ultraviolet Physics group.

Her research on photoionization laser spectroscopy of hydrogen and helium atoms resulted in the experimental proof of a quantum mechanical theory developed in the 1980s. The techniques she developed allowed her to observe for the first time one of the most elusive quantum objects – the wave function. Her paper published in the prestigious journal Physics Review Letter received significant attention in the scientific media all over the world. It was highlighted in the top scientific journals, among them: Nature, Nature Nanotechnology, Science, eEPS, Physics *and* Physics World. *In fact, her work on imaging atomic orbitals was soon chosen as one of* Physics World's Top 10 Breakthroughs of the Year 2013.

In June 2014 she defended her PhD thesis on her groundbreaking technique of photoionization microscopy at the Radbound University in Nijmegen. She currently works at the TNO Institute in The Hague, Netherlands, with the Expertise Group for Nano-Instrumentation.

A few years later in 2019, physicists Zlatko Minev and Michel Devoret at Yale University announced that they had actually observed an electron make a quantum jump between two states within an artificial atom. The artificial atom was created by using wires cooled to superconducting temperatures, which were known to display individual quantum states. If they tried to detect the electrons jumping between two neighbouring quantum states in the wire, they would have disturbed the electron wave function and it would have immediately collapsed into either the one state or the other. Instead, they created a three-state jump where the electron jumped from A to B or A to C. They used a second laser pulse to place the electron in state B and then measured its properties in state B to indirectly measure its properties in state C without making any direct measurement of state C that would interfere with it and collapse its state as viewed by the observer. The transition of the electron from state C to the ground state becomes a protected quantum state that cannot be directly observed. By monitoring this system and counting the rate at which the electron decayed from state B, they could detect the instants when the electron jumped into the state C. What they discovered was that the 'jump' takes about four microseconds but is a gradual process. The electron starts out as 100 per cent in state A, and ends up at 100 per cent in state C, but within the four microseconds time period it enters an evolving mixed state that is for example 90 per cent A and 10 per cent C, then 50 per cent A and 50 per cent C, then 10 per cent A and 90 per cent C before finally arriving at 100 per cent C. This is what would be predicted by the Schrödinger model, which required a smooth process of change from state to state, but with the caveat that, following Heisenberg, there seems to be no sign of the electron in the physical space between the two states.

Quantum physics and determinism

In the late 1980s, investigators working in an area called quantum optics began to develop a scheme for tracking individual photons through optical systems. This led to what became known as Quantum Trajectory Theory (QTT). Photons as quantum objects are not covered by Schrödinger's equation so QTT evolved to provide a mathematical background for predicting single-photon behaviour. This is a very different set of mathematical tools than standard quantum mechanics, because for material particles like electrons there are no such things as specific physical trajectories, only clouds of probability that evolve in time and hide what individual electrons are 'actually' doing. But even so, there is an interesting dodge that lets you track a quantum particle, which was proposed by the Soviet physicists L. Landau and E. Lifshitz in their seminal 1977 book *Quantum Mechanics: Non-Relativistic Theory*. What they proposed is that a particle can be localized to a cubic volume whose size is set by HUP, and these volumes can be strung together to define an average particle path through space. The details of which path a specific particle takes inside this volume 'tube' cannot be known because there is not enough measurement information available to specify the particle's initial position and momentum. With QTT, if you have complete knowledge about the state of a system thanks to dramatic improvements in measurement technology and sensitivity, you can pinpoint the starting state of any system and follow its evolution in time very much like a classical 'baseball' trajectory. Quantum indeterminacy, which through HUP is at the heart of Schrödinger's wave function approach, is just an average over these individual quantum trajectories that is set by our not having better knowledge of a system's physical state. The QTT approach still has random errors in it, but this randomness is not 'baked in' to the theory. It is instead a problem with unavoidable measurement error which can in principle be reduced. For example, environmental influences can introduce errors, but these can be carefully reduced, usually by cooling the system to cryogenic temperatures.

If an atom were perfectly isolated from its environment it would remain in the same quantum state forever. Real atoms, however, are constantly interacting with neighbouring atoms and other aspects of their environment. While the Schrödinger equation tells us about the quantum properties of an isolated atom, QTT describes how a quantum system evolves in time due to its interaction with its environment. Normally we have incomplete knowledge of exactly how an atom interacts with its environment to exchange energy. That is why we end up with an average answer to the question of what a specific photon or quantum particle is doing because we have to average over a large number of possible trajectories to find out what a single typical particle might be doing. But as our knowledge of the exact state of the atom and its environment improves, the number of trajectories we have to average together decreases in number. Eventually, if we have 'perfect knowledge' of the initial state, we need only one trajectory and that is for a single particle. The problem is that this trajectory is not to be thought of as a path through space like a baseball in flight. Instead, it is a path through a series of possible quantum states in what is called Hilbert Space. Normally, this path is broadened by HUP, and defined by solving Schrödinger's equation, but in QTT the path can be made very precise because a detailed knowledge is available for how the system interacted with its environment in the recent past. This is something that Schrödinger's equation can't do.

QTT could not be tested as recently as a decade ago because there were no measurement technologies that could capture more than 90 per cent of the available information on the state of a system interacting with its environment. In 2019, Minev and Devoret not only set up the experiment to catch a 'quantum jump' and follow its evolution in time, but their instruments were able to achieve measurement efficiencies of over 90 per cent so that the uncertainties in the initial states were reduced. This allowed the quantum trajectories of individual electrons during a quantum jump to be resolved. More importantly, with this knowledge in hand, they were even able to reverse the quantum jump literally in mid-flight, and in a completely deterministic way, which means that they were able to avoid the dreaded collapse of the wave function itself.

CHAPTER 9:

Locality and Realism

For all of human history, whenever we looked at a lump of gold or even the sun in the sky we had the distinct impression that we were seeing something that actually existed. It had a permanence to it that didn't matter if it was you doing the looking or someone in the next town. There were some concerns raised by ancient Egyptians and many of the civilizations in the ancient Americas that at least in the case of the sun, some kind of incantation or sacrifices had to be offered each morning to ensure that the sun would in fact rise again each day. But a similar concern never seemed to be raised for other seemingly concrete aspects to our world. In physics, realism is the idea that a particle has specific, intrinsic properties that do not change upon measurement or observation. Locality is the idea that information and physical influences cannot travel faster than the speed of light, so a particle is connected to external events by the light travel time between them. A variety of recent measurements show that quantum mechanics violates both realism and locality, especially through the mechanism of entanglement.

The nature of reality

Philosophers since Parmenides in 5th century BC Greece have spent enormous intellectual time pondering issues of the nature of the physical world and reality itself. Schools have appeared that investigated the simple process of how we gain knowledge about our world from our senses, called epistemology. In this school, *a priori* knowledge is knowledge we gain independently of experiences in the external world through our internal logical reasoning, while *a posteriori* knowledge is gained through sensory experiences in the external world. This led to lines of thinking such as empiricism, which stated that knowledge only comes from sensory experiences guided by logic. The beginnings of epistemology came about in ancient India with the Vaisheshika school in the 5th century BC founded by the philosopher Kanada, and with the ancient Greek Empiricist School in the 3rd century BC led by Serapion of Alexandria. Aristotle's notion that we are all born with a *tabula rasa* 'blank slate' upon which our experiences are written became popular beginning in the Middle Ages in the Arabic world, and then later during the Western Renaissance. Ultimately, these ideas of how we gain knowledge about an external world beyond our own purely mental inner worlds led to the 1920s ideas of logical positivism in Berlin and Vienna in which only statements that could be verified through direct observation are meaningful. All of these philosophical considerations have been transformed through the developing knowledge of the quantum world, where physical reality is now a collection of quantum waves, and the act of learning about their states can alter them irrevocably by collapsing their wave functions. In the past, the observer was separate from the observed, but in the modern world this separation is no longer quite so distinct.

The first inkling that nature was about to present us with a challenge to our ideas of what the external world might be came in the 1800s when it was firmly established from Thomas Young's experiments that light was a wave-like phenomenon. Its wavelength could even be calculated exactly from single- and double-slit interference experiments. This characteristic of light was finally explained by James Clerk Maxwell in his famous set of equations for the electromagnetic field and its waves, which were identical to light waves. Everything seemed to be fine until the early 1900s when Max Planck and Albert Einstein developed the quantum idea for light after deliberation on the way in which light was emitted by heated objects, and the way it produced the photoelectric effect in metals. Neither of these could be explained if light were a wave-like phenomenon. So at this point a dual nature for light had to be accepted. When you set up some experiments in one way, they reveal the wave-like character of light, but if you set them up in another way, they only reveal the quantum character of light. You cannot create one consistent explanation or experiment that covers both of these diametrically opposite conditions: you could test light's wave properties but not at the same time measure its quantum properties and vice versa. Meanwhile, light flitters about the universe going its own way. It is only when a human decides to measure its properties with a specific experiment that light manifests one of two opposite natures. Physicists hardly had time to deal with this paradox when in 1924 de Broglie discovered matter waves and the same dualism was now seen in solid matter itself. This led to Schrödinger's wave mechanics and the idea of the quantum state as a feature of electrons within atomic space.

When I discussed wave functions in Chapter 6 the emphasis was not on their mathematical forms but on what they represented, physically. They were statements of where we expected an electron to be if we measured its location inside an atom, but taking into account their de Broglie wave-like characteristics. In a given quantum state represented by the measurable quantum numbers n, l and m, we could map-out within the atom where a measurement would predict its location after many hundreds or thousands of repeat measurements. But the outcome of a specific measurement in that quantum state could only tell us that at a specific location the electron might only have a probability of say 2.5 per cent to be exactly at that spot. However, all of this uncertainty disappears if we actually detect the electron at a specific location. What seemed to be incomplete is exactly how this act of observing caused the electron described by the wave function Ψ to crystalize around exactly one specific state after the measurement. In no other areas of physics do we have to deal with probability and things like wave functions as almost physically real things in order to describe matter.

THE QUANTUM NATURE OF BASEBALL

A baseball flying through the air is following a path as a definite, visible, solid object that can clearly be tracked along its entire trajectory from place to place with 100 per cent certainty. What the wave function view was telling us is that the simple act of measurement and observation was somehow an important, hidden, intermediate step in acquiring knowledge about a physical event. In our baseball analogy, we calculate all of the possible paths the baseball can take from the batter's impact to the distant point in the outfield where it comes to rest. Each of these paths is a possible 'state' for the baseball within the 'atom' formed by the baseball stadium. We do not know which path the ball will take before the batter hits the ball, although the known abilities of each batter might tell us which particular families of states and trajectories he or she might tend to favour. After the ball is struck, which could be likened to an act of measurement, the specific state of the ball crystallizes around only one possibility. All of this happens in a split-second and we can actually observe the baseball travel its pre-calculated state trajectory. Our further act of observing it in flight has no effect on it remaining in that state until it ends up on the ground. In quantum physics, all of the baseball 'states' are the probabilities for the electron in a variety of possible quantum states, but we cannot individually see the electron in these quantum states any more than we could 'see' all of the possible trajectories that the baseball could have taken. Unlike the baseball, the act of observing the electron disturbs its quantum state, Ψ, causing it to crystallize into one of the possible 'baseball states'. So, between the time of the batter's impact and the ball arriving at a specific place on the ground, the electron can travel a large number of paths to connect these points, and manifest a variety of different quantum states, but we can never know exactly which one it took without making a fatal measurement that disturbs the entire process. During this intermediate time, the electron is in a mixed state. This is where the quantum story becomes interesting.

Collapse of the wave function

Suppose you had a system in an initial state, that could be in either of two final states: A or B. Before you make an observation that confirms which one, your system will be in a mixture of 50 per cent A and 50 per cent B, or perhaps 11.5 per cent A and 88.5 per cent B. But after you make an observation, you discover it is in state B with 100 per cent certainty. For a quantum system such as an electron inside an atom, the fact that it is in a mixed state is not very remarkable. That is just a feature of how strange the quantum world is when it comes to Schrödinger's wave functions. But what Schrödinger himself proposed was a startling experiment you could perform in the world-at-large that had exactly the same features as the quantum world. This hypothetical experiment came to be known as the Schrödinger's Cat experiment.

Place a cat in a box with a mineral that emits radioactive particles with a half-life of one hour so that after one hour there is a 50/50 chance that a radioactive particle was emitted. Now have a Geiger counter register the arrival of one of these particles and then throw a switch that causes a flask of poisonous gas to be released that would instantly kill the cat. Now seal the box so that you could not look inside. After an hour is the cat alive or dead? A strict quantum state approach would say that, before you opened the box to check, the cat was in a 50 per cent alive and 50 per cent dead mixed state. It is the act of you opening the box and looking that 'collapses' the cat's state into 100 per cent one or the other. This paradox of the cat in the mixed state is resolved if you simply remember that the wave function is not a physical object but merely a statement of probabilities for all possible measurement outcomes. With that, this paradox becomes completely trivial and is simply a statement of what you know about the cat. It is a statement of what your level of certain information is. The cat itself has no physical experience of being in a mixed, zombie, state. It will either feel itself to be alive when you open the box, or being dead. It is your information about it that has changed.

Have any of these Schrödinger's Cat mixed states ever been found in nature? In fact they have been, and they form the basis of a whole new technology called quantum computing, which I will discuss in Chapter 15. Although no cat-sized mixed states have ever been achieved, they have been replicated on quantum and atomic-scale systems. For example, in 2010 Aaron O'Connell at U.C. Santa Barbara and his team created a 10-trillion-atom tuning fork only 1 micron wide and 40 microns long and placed it in a chamber cooled to 0.02 kelvins above absolute zero. When set in motion it vibrated at six billion cycles per second. They were able to place it in a superimposed state of vibrating and non-vibrating for about six nanoseconds before its environment finally disturbed this precise mixed state.

The Schrödinger's Cat experiment.

Does an object know what it is?

Another issue related to Schrödinger's Cat is the idea that an object has specific properties even before you measure the object. In this situation, an electron or a photon are in a definite physical state, and the act of measurement only serves to read out what state it is in. Another application is in the area of resolving the particle-wave duality of a quantum system, which is what physicist John Archibald Wheeler attempted to do in 1978 by proposing the Delayed-Choice Experiment. A photon behaving as a light quantum would pass through the experiment in one definite way, but in a completely different way as a wave. Through a clever and rapid switch in the experiment set up, the photon had no particular state until it was actually detected.

The set up for the experiment involved a light source creating a single photon, which would strike a 50 per cent transparent mirror and either pass to one detector or to a second detector along two different paths that crossed at a point inside the experiment. Now a second beam-splitting mirror was placed at this crossing point. This time the photon acts like a wave and seems to split into two, but now these waves interfere with each other and either pass to one detector or the other depending on whether their waves reinforced each other (A+A = 2A) or whether they destroyed each other (A-A = 0). If you insert this second mirror after the photon has already entered the experiment and 'decided' whether to act like a wave or a particle, you can test whether the photon is in a definite state before the measurement, or whether it has no definite state (wave or particle) until the act of measurement. Many of these delayed-choice experiments have been conducted and the results are always the same. Photons and even electrons are not in a definite state of manifesting wave or particle aspects until the moment they are measured. This means that Realism is violated by quantum particles, and objects at least at this scale are not in any definite quantum state until they are measured: they don't even know if they should act like particles or waves and it is the act of measurement that makes this change of reality happen. But this leads to a curious second question: what constitutes a 'measurement'? And how is this information recorded that unalterably changes the state of the particle?

The issue of measurement and local realism cuts to the heart of what role the observer has in the measurement process. Although part of this is treated in the Schrödinger's Cat experiment, the idea that an observer 'participates' in making natural phenomena real in the quantum sense has been a dramatic source of discussion for over 50 years. The conclusion is that no two observers actually share the same reality. This has deep implications for the entire scientific process, which relies on a shared set of facts among many observers. Physicist Eugene Wigner offered a revised Schrödinger's Cat experiment in 1961 to test this idea. It has since been called the Wigner's Friend Experiment.

Wigner's Friend is in a room with an experiment that measures the spin of an electron (up or down), or the polarization state (vertical or horizontal) of a photon. Both of these particles can have a 50/50 chance of being either in the 'up' state or the 'down' state. Wigner's Friend makes the measurement and finds that the particle is definitely in the 'up' state and makes a mental note of the result. Wigner, meanwhile, can look at his friend and see her make the measurement but cannot tell which of the two possible states the particle was in. For Wigner, his friend is now in a tangled quantum state with the experiment such that all Wigner can tell is that his friend either measured one state or another with a 50/50 probability. This is called a quantum superposition. Eventually, Wigner's Friend tears down the experiment and says goodbye to Wigner, but never tells Wigner the outcome of the measurement. Thirty years later at a physics conference, both Wigner and Wigner's Friend are presenting papers on their research when they happen to meet for coffee at a local café. Wigner finally asks his friend 'Oh by the way, what was the spin direction of the particle you measured in my lab 30 years ago?' to which Wigner's Friend replies 'Gee I thought I told you! The spin was in the up direction'. What had happened was that Wigner had experienced his friend in an entangled state with the quantum experiment for 30 years, and only when his friend finally told him what the measured spin state was, did the entangled quantum state finally 'collapse' into a definite state.

For 30 years, Wigner and his Friend had lived in different realities in terms of the quantum measurement. In the Friend's reality the spin had a definite 'up' direction, and she carried on her professional career using that factum in her own research, but in Wigner's reality it was still in a superposition of 50 per cent up and 50 per cent down and his research based on that experiment evolved in a very different direction. What is disturbing is that Wigner's Friend has been living in an entire universe (out to 30 light years in fact) where the quantum state had collapsed to one answer, while at the same time Wigner's entire universe (out to 30 light years) has been in the superimposed state where both outcomes were still possibilities. This is a clever thought experiment, but does it really reflect the real world? It seems that it does.

Locality and action at a distance

Another feature of quantum states related to the Schrödinger's Cat effect is what physicists call Locality. This is the basic principle that objects are only affected by influences that travel at the speed of light or slower. Einstein's relativity theory is based on the premise that no information, energy or matter can travel faster than the speed of light in a vacuum (299,792 km/s), so if Object A is influencing Object B located one kilometre away, the influence cannot arrive in a time shorter than the light travel time between them or $\frac{1}{299792}$ seconds. Without allowing for this effect with high precision, our entire network of Global Positioning System satellites would never work to give us ground measurements more accurate than 10 km without daily resets of the clock synchronization. Other examples include the 8.5-minute delay for light to travel from the sun to earth. Any major solar flare producing a burst of x-rays on the sun would take 8.5 minutes for its electromagnetic effects to reach earth and upset its ionosphere. Quantum mechanics seems to upset this immutable relativistic rule. Einstein did not like the Schrödinger's Cat idea of mixed and collapsing quantum states. He believed, for this reason, that quantum mechanics was still incomplete. To argue this point with an analogy, in 1934 he, Boris Podolsky and Nathan Rosen came up with a hypothetical experiment, called the EPR Experiment, that physicists ought to be able to perform but could not at that time.

The first step is to create two photons travelling in opposite directions and with opposite polarizations. The easiest way to do this is to collide an electron with its antimatter twin, which will produce two photons travelling in opposite directions. Another method is to wait for a sub-atomic particle called a neutral pi-meson, ϖ^{0}, to decay into two photons. Other modern methods use lasers to pump electrons into atomic quantum states that produce pairs of photons when the states decay in 'double-photon emission'. Mathematically, these two photons are in a single quantum state rather than two separate ones, which is a phenomenon called entanglement. The reason for this is in the next step.

BEYOND THE THOUGHT EXPERIMENT

In 2019, physicist Massimiliano Proietti at Heriot-Watt University in Edinburgh and a few colleagues performed the Wigner's Friend experiment using six entangled photons, which created two different realities that coexisted although they were irreconcilable. The experiment details are complex but it created the Wigner's Friend measurement and then tested whether from the Wigner perspective that Wigner's Friend was still in a superposition with the photon spin measurement. The system involved a photon whose spin was measured and recorded in the polarization state of another photon representing Wigner's Friend. This completes the act of measurement by the Wigner's Friend part of the experiment, which is the first piece of information. The original photon is then detected and recorded as a completed measurement, which is the second piece of information. Finally, a third photon representing Wigner interacts with the Wigner's Friend measurements. The result is that the Wigner system records the definite detection of the original photon, but remains in an entangled state with the information stored by the Wigner's Friend photon. Because these two realities are incompatible, the original conclusion seems to stand that in the quantum world multiple observers share multiple different realities for specific measurements that may not be compatible.

An additional quantum property of photons and matter is that they both behave as though they have a quality called spin, which I will describe in the next chapter. For photons, this spin quality translates into the photon having a unique attribute called circular polarization. If the pair of photons is produced in an entangled state, one photon will be emitted with a right-handed polarization (denoted as +1), the other must have a left-hand polarization (denoted as -1) so that the sum of the two is zero, matching the net spin of the source of the photons. Electrons can also be produced in entangled pair states that have zero net spin and so one electron has to be 'spinning in the up direction' and the other has to be 'spinning in the down direction' so that the sum is again zero. The result of the spin constraint for the entangled states is that if you measure one of the photons to be in +1 spin state, you know immediately that the other photon has to be in the -1 spin state without even having to measure the second photon. The same situation applies to electrons in an entangled state. If one is measured to be spin-up, the other electron has to be in a spin-down state. The difficulty occurs when you view this measurement process in detail.

When you make the measurement on one of the particles to identify its specific spin, the other particle has to seemingly adjust itself to display the opposite spin. What this means is that if you have two observers at the locations of A and B measuring the spin of each particle, as soon as the A observer measures the particle to be in the +1 state, the B observer will measure the particle there to have the opposite -1 spin. The distance between these particles can be made long enough that an information signal travelling at the speed of light will not make it to observer B before that measurement is made. There are two possible solutions to this paradox, either quantum information even at the macroscopic scale can travel faster than the speed of light, which is not allowed by relativity theory, or the correct information is already present in particle B because there is hidden information outside of quantum mechanics. Einstein preferred this 'hidden variables' idea as his solution to this problem. A series of experiments over the last 20 years conclusively demonstrates a very different interpretation.

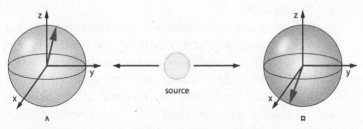

Two particles A and B in an entangled state have equal and opposite spins.

Bell's Inequality and Local Realism

In 1964 John Bell at the Institute for Advanced Study in Princeton, New Jersey, offered a groundbreaking test to confirm or disprove the EPR effect. Known as 'Bell's Inequality' it tests the measurement outcomes for the correlations in the A and B spins and determines whether the statistical results are either consistent or inconsistent with the predictions of quantum mechanics. Detailed experimental studies were conducted in 1972 by American physicists John Clauser and Stuart Freedman, and later in 1981 by French physicist Alain Aspect and his team. Both studies confirmed the predictions of quantum mechanics and that quantum entanglement is a reality that is inconsistent with any hidden-variable theory. This means that the apparent fact that information travels faster than light in an entangled state of two particles cannot be eliminated by creating a better theory beyond quantum mechanics that provides the required missing information at the location of the measurement. So, if quantum mechanics is correct, the members of an entangled state seem to exchange information at faster-than-light speeds; however, this is an illusion because the two observers cannot use this to communicate with each other. The correlation only appears in a statistical sense over many trials. Quantum mechanics violates the principle of locality, and there is no cure for this by creating a super-theory. In other words, quantum mechanics is essentially a complete theory and this non-local behaviour is a true feature of our physical world.

So the bottom line is that quantum mechanics presents to us a world in which no two observers share the same reality, that the theory can offer the possibility of certain kinds of information travelling faster than light (Locality) and that a quantum particle is not in a definite state until the moment of measurement (Reality). In short, quantum theory is a non-Local, non-Real theory that nevertheless seems to reflect how nature works at the quantum scale and even at the macroscopic scale for certain kinds of information. Amazingly, quantum mechanics can be either non-Local or non-Real but not both at the same time. In 2016, physicist Peng Xue, at Southeast University in Nanjing, China, used entangled photons to create a superposition of three quantum states. By performing various measurements on these photons, the researchers could violate Locality and Realism separately, but not at the same time.

JOHN BELL

John Stewart Bell (1928–90) was born in Belfast, Northern Ireland. At the age of 11 he decided he wanted to become a scientist and following his secondary school education at the age of 16, he enrolled at the Queen's University of Belfast and received degrees in physics and mathematics. He then received his Doctorate from the University of Birmingham in 1956 where he worked on problems in nuclear physics and quantum field theory.

While working at the accelerator facility at CERN, Bell became interested in the old pilot wave theory by de Broglie and also the Einstein-Podolsky-Rosen Paradox, which led him to investigate the mathematics of wave function collapse and how to test whether hidden variables exist. This resulted in his pioneering paper in 1964 where he proposed a simple 'inequality' equation that observations of entangled quantum states have to obey.

This Bell's Theorem also became a sensitive test of whether quantum mechanics was incomplete and needed to be improved by hidden variables. He was heavily influenced by Davis Bohm's nonlocal hidden variable theory, and his theorem was put to the test beginning in 1972 by Alain Aspect and others. His wife Mary Ross was also a physicist and they collaborated on many projects involving the design of particle accelerators. Bell died in 1990 from a cerebral haemorrhage, and was on a shortlist of candidates for the 1990 Nobel Prize in Physics for his many contributions to the deep understanding of quantum phenomena.

John Stewart Bell.

CHAPTER 10:

Relativity and Antimatter

The previous chapters have provided a virtual fire-hose of new ideas about the nature of matter and the many bizarre ways that the atomic world operates. The neat, trim rows of fuzzy atomic balls that we see in images from powerful microscopes belie a peculiar universe within the atom that challenges just about every precept we have about how the world operates. We can't even be sure that two different observers are experiencing and recording the same information, or that objects have well-defined properties before they are observed and measured. When Albert Einstein's theory of relativity is added to quantum mechanics, this relativistic theory predicts the existence of a twin to ordinary matter called antimatter. This new occupant of our universe can be found in a variety of settings, but when in contact with its matter twin, the two annihilate completely and produce a burst of pure energy.

Relativity and quantum mechanics

At the same time Schrödinger and Heisenberg gave us the mathematical means to make predictions about the quantum world, another avenue was opening up that forced the entire edifice of quantum theory to be re-written. During the formative years of quantum mechanics beginning with Bohr's atomic theory, Albert Einstein announced a complete restructuring of Newton's universe in 1905 and 1915 around the principles of relativity. There were two key *a priori* assumptions that had to be fully obeyed in his special theory of relativity. First, observers moving at constant speeds relative to each other have to agree about the physical laws they experience and discover. This is known as the principle of covariance. The second assumption is that no particles, energy or influences can travel faster than the speed of light in a vacuum.

In the 1800s the only way of making this speed adjustment, called a Galilean transformation, was to use the physics of Newton as a guide. In Newton's physics there was a simple way to make this change, but when you applied it to what your friend on the pavement should see with the currents and electromagnet the answer is still wrong (see box opposite). What Einstein's special relativity provided was the correct coordinate transformation so that Maxwell's electrodynamics worked at very high speeds. It also led to a similar application by extending Newton's physics so that they are now correct for high-speed motion as well. The key ingredient to this correct transformation, called a Lorentz transformation, was the assumption that no motions may exceed the speed of light. Einstein had found the reason why the Lorentz transformation worked because it was now explainable as a direct consequence of the speed of light being the fastest speed for matter and energy. Special relativity also did something else. When you re-wrote Newton's or Maxwell's equations so that they obeyed the Lorentz transformation, the equations looked identical in form. This is the statement of covariance, which means observers in different reference frames will record the same physical laws if they are in constant relative motion. In his 1905 paper *The Electrodynamics of Moving Bodies*, Einstein showed how his two principles led to the unification of Maxwell's theory of light and electrodynamics, with Newton's successful mechanics of moving bodies. Now an entirely new theory of nature had appeared in the form of quantum mechanics, and the race was on to re-write this theory so it was consistent with special relativity. One would think that such a difficult task would have taken decades, but in fact it was accomplished less than two years after Schrödinger published his theory of wave mechanics.

RELATIVITY IN ACTION

Imagine yourself in the following situation. You are travelling at high speed inside a train. For some reason, you are experimenting with currents flowing in a wire connected to a battery in order to create a toy electromagnet. Nothing unusual happens, and you can use the equations developed by James Clerk Maxwell in the 19th century to predict exactly how much current will produce a specific amount of magnetism. Now suppose you are on your smartphone describing this experiment to your friend standing on the platform as the train passes by. She texts you that she is seeing a different current flowing in your wires and a different strength for the electromagnet. If she were to write down Maxwell's equations based on what she was observing in your reference frame she would get equations that look a bit different than yours. How could this be? Shouldn't there be one set of physical laws that govern currents and magnetic fields? It's because the motion of your car is making the current appear stronger or weaker to your friend depending on which way it was flowing relative to the direction of the car's motion. OK, but there should still be a way of accounting for this velocity difference and getting the right answer for what your friend on the pavement seems to be seeing.

For the observer by the trackside time passes faster than for the observer on the train.

The English theoretical physicist Paul Dirac was already deeply involved in quantum theory having read the work by Heisenberg in 1925 and completed his PhD dissertation on the mathematical aspects of this theory by the following year. Also a few years earlier the Austrian physicist Wolfgang Pauli had proposed a new quantum number for electrons called 'spin' and that electrons would have two spin states, 'up' or 'down', each with a magnitude equal to ½ h. The spin quantum number s would be in addition to the three other quantum numbers n, l and m already known for atomic states. This property of quantum spin was also important for photons, which the Indian physicist Satyendra Nath Bose had discovered in 1924–25. However, photons carry 1 h-unit of spin and so behaved very differently than the electrons. What Dirac did was to take into account the spin characteristics of electrons and quickly write down what the form of Schrödinger's equation should look like if it was fully consistent with special relativity. Some say he received much of his inspiration for discovering the form of this equation while staring into a fireplace in Cambridge, England, where he was living at the time. This new relativistic form of Schrödinger's equation tremendously improved the accuracy of predicting exactly what electron quantum states should look like, but they opened the door on an entirely new universe called antimatter.

PAUL DIRAC

Paul Adrien Maurice Dirac (1902–84) was born in Bristol, England, and was known throughout his life as being taciturn to the point that his colleagues defined a unit called 'the Dirac' as one word per hour.

Following his formal education he went on to study electrical engineering at the University of Bristol. In 1923, having accumulated enough money to live in Cambridge, he enrolled at Saint John's College where he studied Einstein's general relativity, and in 1926 he submitted the first PhD thesis on the new theory of quantum mechanics.

Dirac's brilliance in theory was legendary. Apparently, in working to create a theory of quantum mechanics that was consistent with relativity, Dirac by a fireplace one evening is alleged to have written down with no rough drafts the exact relativistic version of Schrödinger's Equation.

At the Solvay Conference in 1927 a session discussed religion and science to which Dirac noted 'I cannot

Paul Dirac.

*understand why we idle discussing religion. If we are honest
– and scientists have to be – we must admit that religion is a
jumble of false assertions, with no basis in reality. The very
idea of God is a product of the human imagination.' To
this, Werner Heisenberg noted 'There is no God and Dirac
is its Prophet', to which Dirac and the audience burst into
laughter.*

*Dirac died in 1984 and although buried in the Roselawn
Cemetry in Tallahassee, Florida, a commemorative marker
to him can be found in Westminster Abbey and inscribed
with Dirac's equation.*

The Dirac Sea and its denizens

At the core of Dirac's relativistic theory was a key feature of
relativity; namely, the idea of an invariant quantity that all
observers would agree upon regardless of their state of motion
at high speeds. To see how this works, take a metre stick and
write '1 metre' on it. Now while your friend watches, move this
metre stick around so that your friend sees it appearing in many
different degrees of foreshortening. From your perspective, its
length is still exactly one metre. From your friend's perspective,
and knowing its length is exactly one metre, she can calculate
what its orientation is in space. In special relativity, the total
energy of an object defined by $E^2 = p^2c^2 + m^2c^4$ serves the same
role as the metre stick where 'p' is the momentum of the particle
in motion, m is the mass of the particle and c is the speed of
light. Notice that when the particle is not moving, p = 0 and we
get Einstein's famous equation $E = mc^2$. But as any secondary
school student knows, if $E^2 = a$, then there are two roots $E(+) =
+a$ and $E(-) = -a$. For all familiar physical situations, we ignore
$E(-)$ and report only the positive root for a particle's energy.
In the quantum world, however, this is not possible. Instead,
Dirac's equation allows particles to have two possible energies

$E(+)$ and $E(-)$. This turns out to be a big problem, because if the momentum of the particle in an observer's reference frame were zero because the observer is moving along with the particle, then you have $E(+) = +mc^2$ and $E(-) = -mc^2$. Since the negative sign cannot come from the c^2 factor, this means there are two kinds of matter: positive matter (+m) and negative matter (-m). Dirac was troubled by the implications of these apparently non-physical negative-energy states. Between 1928, when Dirac published his equation, and 1930, the paradox reigned over the unfailing accuracy of Dirac's equation which was its major success, and its failings in predicting nonsensical $E(-)$ states for matter that had no physical explanation. In fact, it was recognized that if these energy states existed at all, all material particles with $E(+)$ would immediately cascade into these lower-energy states and once again, as for the Bohr model, all matter would become unstable. Then in 1930, Dirac himself proposed a way out of this dilemma: the negative-energy sea.

What if these $E(-)$ states were already occupied? Then there was no place for the normal matter $E(+)$ states to go. Dirac proposed that, using the Pauli Exclusion Principle, these particles would have identical quantum states to normal matter so that you could not put two of them into the same negative-energy state, which allowed the $E(+)$ states to remain stable. In his paper, he also noted that a missing negative-energy particle, called a 'hole' would resemble a normal matter particle with the opposite charge and a positive energy $E(+)$. At that time, the only such particle known was the proton, so in one fell swoop Dirac had provided an explanation for why protons exist inside atoms. But the mathematician Hermann Weyl had another idea. Arguing from symmetry, he proposed in 1931 that the new hole particle had to be some new kind of particle with the mass of the electron but with opposite electric charge.

Around the same time, Caltech physicist Carl Anderson, who had been studying cosmic rays using the new invention of the bubble chamber, had just announced his discovery of what he

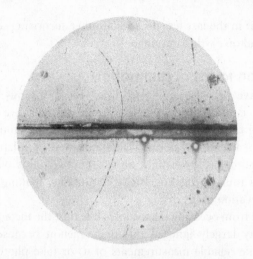

Anderson's original cloud chamber photo of a 63 MeV positron travelling upwards, striking a lead plate and continuing as a 23 MeV particle. From the curvature the particle had a positive charge, and from the energy loss its mass was calculated to be equal to an electron.

called the *positives*, later known as the *positron* or 'anti-electron'. With that, Dirac's prediction of a sea of negative-energy particles and holes was experimentally vindicated.

Apart from cosmic rays and super-expensive research-grade equipment, are there any other places in our familiar world where antimatter raises its head? At the level of individual particles they are, of course, too minute to be seen, but when enough of them get together they can collectively produce some dramatic fireworks. The reason for this is that when a single electron and positron collide, they have to release the full $E = mc^2$ of each of their masses into two photons of light energy. Each of these photons has a wavelength set by Planck's Law, so you have $h\nu = mc^2$. With an electron mass equal to 9.1×10^{-31} kg, you get a photon energy of 511 keV and a frequency of $\nu = 1.2 \times 10^{20}$ Hz. Light at this frequency is called gamma radiation. So where do we find radiation at this frequency? Usually inside nuclear power plants, but in the last few decades another surprising source has been found on earth: lightning.

Common forms of antimatter

People have been observing lightning for thousands of years. However, careful observers over the centuries have reported something odd happening in the space above a thunder cloud. In 1730, a keen-eyed German amateur scientist Johann Georg Estor noticed flashes of light above thunderclouds while on a geological tour of the Vogelsberg mountains, making notes of this observation in his diary.

Apart from occasional anecdotes like this, the meteorological community largely ignored this phenomenon because no one could make reliable measurements of it, or take photos of this effect. So research on these flashes languished for another 100 years. It took this long for high-speed photographic technology to catch up with these mercurial millisecond flashes. On 6 July 1989, Robert Franz at the University of Minnesota accidently captured a video record of sprites while surveying the night sky for aurora and meteors. The same year, NASA's STS-34 astronauts captured multiple images of these events above thunderstorms over northern Australia. Since then, these events and many others have been extensively recorded and analysed, revealing an entire bestiary of 'sprites', 'elves', 'jellyfish', 'blue jets' and 'tendrils', each with its own role to play in upper-atmosphere, transient luminous events. This complex, electrified environment appears to be the breeding ground for antimatter.

The intense electric fields at the tops of thunderstorm clouds can accelerate electrons to such high energies that, when they collide with air molecules they release bursts of gamma rays. Some of these collide with atomic nuclei and produce beams of electron-positron pairs, which quickly annihilate and produce a second type of gamma ray light at an energy of 511 keV. It is this radiation that NASA's Fermi Observatory can detect from space, and is a direct and unique fingerprint signature of antimatter production and annihilation. During one of these events on 14 December 2009, Fermi was passing over Egypt and detected the gamma ray burst produced by a thunderstorm over Zambia over 4,500 km away. Over 130 of these events have been recorded during a three-year period, although millions of these intense lightning events probably occur each year around the world.

ANTIMATTER AND MEDICINE

Another natural source of antimatter is in medical diagnostic imaging. During a procedure called Positron Emission Tomography (PET), a patient is injected with a tracer, often fluorodeoxyglucose, whose fluorine atoms have been switched with their radioactive isotope Fluorine-18. This isotope emits positrons in its 107-minute decay about 97 per cent of the time. Tissues take up this form of glucose as part of their metabolism and when the positrons are emitted, they immediately are annihilated, producing two gamma rays that quantum mechanics requires to move in exactly opposite directions so that the total spin is zero. Detectors register these pairs of gamma rays and can work backwards to the point in the tissue where they were emitted, forming an image over time. Cancer cells have higher metabolisms than normal cells, so normal tissue remains dark while cancerous tissues 'light up'. PET scans are also used to record brain activity during functional studies of how brains process information, and also indicate various forms of dementia and psychological conditions.

The addition of antimatter into our collection of known types of atomic matter set the stage for many new discoveries that went well beyond merely understanding the nature of atomic quantum states with higher precision. There was now an entirely new spectrum of matter to investigate along with the quantum properties of atoms made entirely out of antimatter. The first antimatter hydrogen atom consisting of an anti-proton and a positron was created in 1995 by a team of physicists at CERN, but these 'hot' hydrogen atoms were moving too fast to study in detail. By 2000, CERN had built the Antiproton Decelerator (AD), which in 2002 produced over 50,000 slow-moving and very 'cold' hydrogen atoms for further study, leading to the first confinement of 38 antihydrogen atoms in 2010. This allowed a detailed study of their quantum transitions, which should be identical to what was found in ordinary hydrogen atoms. By 2017 a team of investigators at the CERN AD lab led by M. Ahmadi made measurements of a specific quantum jump called a hyperfine transition, which was predicted by Dirac's atomic theory, and is the workhorse '21-centimetre' line in astronomy to map interstellar hydrogen gas in the Milky Way. In ordinary hydrogen atoms, its frequency is known very precisely as 1420.405751 megahertz (MHz). What the CERN AD team did was to measure this hyperfine line for 194 antihydrogen atoms as 1420.4+/-0.5 MHz; a precision of about 4 parts in 10,000. This would not be considered a very good measurement for ordinary atomic physics, but for antimatter physics it is currently the state-of-the-art. The bottom line is that first of all, antihydrogen also has the same hyperfine line, and that its frequency is apparently identical to ordinary hydrogen based on measurements so far.

Antimatter and gravity

Another interesting set of measurements involves testing whether antihydrogen falls at the same rate as ordinary hydrogen under the influence of gravity: 9.8 metres/sec^2. This is a very difficult experiment, but teams of investigators at the CERN AD lab using the GBAR and ALPHA instruments are hopeful that soon they will be able to make the measurement. One of the first results should be whether antimatter will fall up or down compared to ordinary matter. No theories predict that antimatter should fall upwards, but this possibility has to be eliminated experimentally

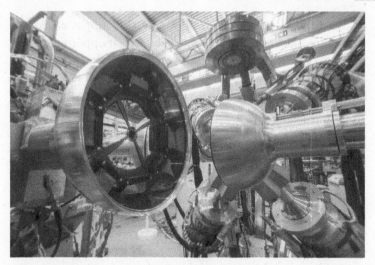

The CERN antiproton Unstable Matter Annihilation experiment.

before more detailed understandings of antimatter can proceed. As an example of how technically difficult this measurement will be, only in 1999 was a freefall test performed on ordinary hydrogen atoms by physicist Achim Peters and his team at Stanford University using an atom interferometer. They found that hydrogen atoms fall with precisely the same acceleration as expected near earth's surface to a precision of three parts in one billion. With antihydrogen, there will be far fewer atoms to test and these will be highly reactive with the ordinary matter in the measurement equipment.

One last, but very important, issue involving antimatter is that it is not detected in large quantities in the astronomical universe. When matter and antimatter come into contact, they produce brilliant annihilation lines at 511 keV and for proton-antiproton collisions the line is at 938 MeV. Because matter and antimatter should make up equal shares of the matter in the universe, these annihilation lines should be among the brightest spectral lines detectable across the universe in many different systems, especially among colliding galaxies. Also, within clusters of galaxies, all of the member galaxies travel through a dilute gas called the intracluster medium. There should be plenty of opportunities for these galaxies to interact with the gas to produce the characteristic annihilation lines for positrons and antiprotons. In fact, none of this radiation has ever been detected, placing a limit to antimatter in our universe at about one part per billion of ordinary matter. Within the stars and interstellar gas of our own Milky Way, less than one part in 1,000 trillion is antimatter, mostly produced in exotic collisions between interstellar matter and cosmic rays. Even though the equations of quantum theory seem to be entirely symmetric when it comes to matter and antimatter, our universe is definitely not, so there must be some mechanism at the atomic scale that breaks this symmetry. In fact, because the Big Bang was a seething dense plasma of elementary particles at the dawn of the history of the universe, this mechanism, whatever it was, came into play when the universe was first formed. This remains one of the most outstanding questions in physics and cosmology even at the start of the 21st century.

CHAPTER 11:

Virtual Particles and the Void

Human fascination with 'invisible agents that control our world' has a long lineage stretching back beyond the dawn of recorded history, but can still be found in the various spiritual movements of the 1800s. Whether it is mediums communing with departed spirits via seances, or spiritualists appropriating concepts of the fourth dimension into their supernatural architecture, we have had an abiding curiosity about hidden worlds and their phenomena. With the advent of the discovery of antimatter, some of these same suspicions now seem to have come to roost in 20th and 21st century physics. The same mathematics used to predict antimatter and make quantum mechanics compatible with Einstein's relativity also forces us to accept that 'empty space' is far from being devoid of matter. Heisenberg's Uncertainty Principle allows particles and phenomena to flourish in the 'empty' void beyond our ability to ever directly observe them. However, quantum particles such as the electrons we can measure are influenced by these hidden processes and provide indirect evidence that they exist. There are, in fact, hidden agents controlling aspects of our physical world, but they are nothing like what millennia of speculation ever prepared us for.

Virtual particles and nature's invisible confederates

From the first rudimentary studies of atomic chemistry, through the advent of Dirac's relativistic theory of quantum mechanics, it seems that we have steadily made progress on this climb. There have been new ideas we have had to learn or simply accept as facts about the quantum universe. Among these was the idea that matter can behave as a wave but without anything physical that actually waves. What we define as 'reality' is observer-dependent, and even the physical properties of the elementary particles we study and measure are not realized until the instant they are measured. Dirac's relativistic theory scored huge successes in its ability to predict how atoms emit light, but as part of the pill we had to swallow was that, not only does antimatter exist, but there is an apparently invisible 'sea' of particles hidden within the fabric of empty space. They are not real particles because they have literally less than no energy at all unlike real particles that have more energy than nothing. It is like sticking a thermometer into empty space that registers temperatures all the way down to absolute zero, and finding that negative temperatures also exist. Since material particles cannot have less than zero energy because their very own masses keep their energies above the positive amount specified by $E = mc^2$, we still have to make peace at least theoretically with these negative-energy particles. The way that physicists do this is to first give them a proper name – virtual particles – but we don't have to reach for our textbook on relativistic quantum theory to get at least a visceral understanding of what they are.

For millennia, humans have been convinced by one argument or another that there is literally more to the world than meets the eye. Behind all of the activity we see and experience, there are invisible agencies that sometimes reach into our world to steer it towards certain inscrutable ends. More than 90 per cent of all humans deeply believe the existence of a supernatural world and make daily or weekly acknowledgements to this idea via rituals and incantations in one form or another. Although there isn't much that can be counted on as proof of such a parallel world, nevertheless, it cannot be decisively disproven either. With 20th century physics, some aspects of such a background world beyond Newton's clockwork universe seem required and inescapable.

The hidden fabric of the world

The French Astronomer Camille Flammarion captured some of these mysterious behind-the-scenes machinations of our world in his famous wood cut from 1888. Another, more humorous example comes from the 1938 movie *The Wizard of Oz* where the Wizard is unveiled from behind a curtain by Toto. Interestingly enough, in the decade before the relativity and quantum revolutions, some physicists where whistling past the graveyard concerned about this same issue. Heinrich Hertz offered this rather sobering comment on the state of physics in the last decades of the 19th century: 'If we wish to obtain an image of the universe which shall be well-rounded, complete and conformable to law, we have to presuppose, behind the things we see, other, invisible things – to imagine confederates concealed beyond the limits to our senses'. To set the proper context, during the late 1800s there was a huge public interest in the supernatural world. This isn't to suggest that scientists of this age were reflecting these popular beliefs in their research, but the coincidence is rather interesting to investigate.

We know that for as long as humans have been on this planet, records show a nearly universal belief that there is a

An 1888 woodcut by Camille Flammarion depicting the idea of a world beyond our physical world with its own phenomena and occupants.

hidden side to the world. Lacking modern instruments that extend our senses, this hidden world was largely relegated to the imagination and populated by gods, spirits, ghosts and all manner of invisible agents. These act in most cases by throwing lightning bolts, sending plagues to torment us, or occasionally casting a rainbow in an unexpected corner of the landscape. With a little imagination, they can even be heard speaking to us in the curious cacophony of a babbling brook. Typically these invisible agents act by altering the probabilities of certain outcomes. This is why, even to this day, some human groups sacrifice animals to help bias random events towards more favourable outcomes by currying favour with supernatural powers who seem to have an insatiable appetite for blood and viscera. More than half of all humans profess to believe that ghosts exist, and that some people have supernatural powers making them 'sensitive' to this hidden world. Astrology is entirely based on the premise that the locations of the sun and planets play a non-gravitational role in making you who you are, even at the moment of birth. Although these beliefs were commonplace for most of human history, the late 1800s seemed to be a particularly aromatic cauldron for cooking these beliefs into a more modern stew of 20th century ideas.

SPIRITUALISM

The Spiritualism Movement began in America c. 1840 with the belief that spirits existed, had advanced knowledge, and could be communicated with to answer a variety of personal questions. This idea was advanced by the self-styled clairvoyant Emanuel Swedenborg and hypnotist Franz Mesmer. Living 'mediums' consulted in trances with spirit guides during events called seances, with the communication facilitated by such technologies as the ouija board (c. 1886) and the talking horn. A huge interest in spirit communications occurred during the American Civil War when numerous families wanted to hear from deceased soldiers. Around the same time, mathematicians such as Bernhard Riemann and Clifford Will in the 1860s were investigating spaces of more than three dimensions to generalize Euclid's geometric investigations beyond flat three-dimensional space. This led to an upsurge in interest in the 'fourth dimension', especially through the writings of science popularizers such as Victorian schoolmaster Edwin Abbott Abbott who published his famous book Flatland *in 1884; the story of creatures living in a two-dimensional world, and communicating with beings from the third dimension. Soon after Abbott's book appeared, H.G. Wells wrote* The Time Machine *in which the fourth dimension became identified with time itself, not an independent spatial direction. This provided even more fodder for Spiritualists.*

The notion that spirits hailed from some other dimension can actually be found in the works by philosopher Henry More who in 1671 proposed that since spirits must occupy space, they could only do so if they existed in the fourth dimension. But it was astronomer Johan Zöllner from the University of Leipzig who really popularized this idea. He was an enthusiastic champion of the American medium Henry Slade's assertion along these same lines. Zöllner's book *Transcendental Physics* of 1878 purported to give experimental evidence of the existence of a four-dimensional spirit world. Edwin Abbott Abbott (see box above) was influenced by these speculations, as was the Christian spiritualist A.T. Schofield in his book *Another World*. Even as late as 1908, P.D. Ouspensky wrote in his essay *The Fourth Dimension* that all living things exist in the fourth dimension, not just spirits, and that the greater part of our being exists in this dimension, only our collective consciousness are not aware of this. Ultimately the Spiritualism Movement became discredited as more and more mediums were found guilty of charlatanism and deceitful practices by the end of the 1800s. Nevertheless, the idea that spirits and disembodied ghosts exist and can haunt people and buildings to influence their behaviour remains a popular belief. It is a belief greatly encouraged by numerous 'scientific' programmes on TV and the Internet that investigate these claims using high-tech equipment and a narrative patois that resembles science but merely uses the trappings of science for essentially entertainment purposes. So even today, the idea that there is an invisible world capable of influencing the behaviour of matter (i.e. people) is a mainstream idea in the public arena. The discovery of such a world as a necessity for science might not be as surprising as it seems within the scope of ideas entertained by people over the centuries. The fact that the discovery literally popped out of the mathematics rather than a scientific instrument is even more remarkable.

Quantum electrodynamics

At about the same time as renewed attention was being directed towards developing a true quantum theory of light, another viewpoint on the subject of how atoms emit light made its way into the literature in 1924. Niels Bohr, Hans Kramers and John Slater made a bold proposal that completely rejected the quantum character of light. In a last bid for a purely classical description guided by Newton and Maxwell, they believed as Maxwell and many others had that light was a strictly continuous phenomenon. Any quantum characteristics it might seem to have had had to do with the way that light interacted with matter. They conjectured that energy conservation doesn't strictly hold for individual atomic events, but is only valid in some average sense when many interactions are added. Something which they called a virtual radiation field was associated with each atom. It contained photons carrying every possible energy difference among the energy states in each atom. If an atom could produce a

Niels Bohr.

spectral line at 6519 Angstroms, there would be a virtual photon with exactly this required energy just waiting to be born under the right conditions. This model was eventually disproved by experiments in 1925 carried out by Walther Bothe and Hans Geiger.

Not long after Dirac discovered the positron and the properties of empty space, he uncovered yet another ingredient to the vacuum state from among his theoretical deliberations. In addition to the great sea of negative energy states and holes, Dirac proposed that the physical vacuum was also a rich cauldron of virtual electron-positron pairs continuously appearing and disappearing everywhere in space in strict accordance with Heisenberg's Uncertainty Principle. The first theoreticians to add these pairs into the mathematical stew of calculating the electron's mass were Wendell Furry and J. Robert Oppenheimer in 1934, and a few months later Wolfgang Pauli and Victor Weisskopf. They discovered in their mathematical deliberations an interesting and unexpected result: the pairs that appeared and disappeared in the vacuum could contribute a slightly negative energy to the electromagnetic field. What the pairs did was to produce an effect called vacuum polarization in the space surrounding the electron. The positive charge on the virtual positron in the pair is attracted by the negative electron, while the virtual electron's negative charge is repelled by the electron. From a distance, the virtual positron cancels out some of the electron's negative charge. The resulting strength of the electron's field would weaken at small enough distances. Can these virtual pairs be detected? Amazingly enough, yes.

According to Dirac's relativistic quantum theory, the lowest quantum state in the hydrogen atom is $1S_{1/2}$, with the quantum numbers $n = 1$, $l = 0$ (S) and $s = \frac{1}{2}$. The next highest with $n = 2$ is represented by $2S_{1/2}$ and $2P_{1/2}$, they have the quantum numbers $n = 2$, $l = 0$ (S), $l = 1$ (P) and the electron spin quantum number $s = \frac{1}{2}$. Because the Bohr model does not include electron spin, the Bohr model would predict that the 2S and 2P states would have the same energy. Dirac showed that with electron spin included, there were actually two of these 2P states with slightly different energies depending on how the electron spin is aligned. If it is aligned 'down' the state is $2P_{1/2}$, but if it is aligned 'up' it is a new state called $2P_{1/2}$. The Lyman-alpha spectral line is due to an electron falling from the n = 2 to the n = 1 energy levels which normally is an energy difference of 10.2 eV yielding a single line at a wavelength of 121.6 nm. Dirac's calculation predicted that this line should be split into two to produce 'fine structure'. The difference in energy between them is about 0.000045 eV. This splitting is very hard to detect, but other lines will be split as well and these are easier to observe, spectroscopically.

An interesting thing happens when you add in the effects of virtual electron-positron pairs popping in and out of the vacuum. Some of the positive charge of the proton is screened, making its electric force on the electron feel less intense. This causes the electron orbitals to move slightly farther away from the proton. Instead of the 2S and 2P states having the same energy (i.e. n = 2), they have slightly different energies by an amount of only 0.0000044 eV or about 10 times smaller than the Dirac 'fine structure' shift. In 1947, Willis Lamb measured this effect, and its only explanation was the effect of this vacuum polarization caused by the virtual electron-positron pairs. There is another way to detect this effect, this time on electromagnetic waves.

When you apply a strong magnetic field to a region of space occupied by matter (a crystal), electromagnetic waves entering this region will have their polarization changed depending on how they vibrate relative to the crystal's lattice structure. This is called optical birefringence. Another way to look at this is that the refractive index of the crystal depends on the way that the incoming light is polarized. It is believed that the Vikings used this effect with calcite crystals to locate the sun in the sky even when the sky was cloudy. With the virtual vacuum, even empty space has this property, but to make it happen you need very intense sources of light and exceptionally strong magnetic fields.

WOLFGANG PAULI

Wolfgang Ernst Pauli (1900–58) was born in Vienna, Austria, and attended the Döblinger Gymnasium where he graduated in 1918. He attended the Ludwig Maximilian University and studied under Arnold Sommerfeld, receiving his PhD in 1921 for his work on the quantum theory of the diatomic hydrogen atom.

Within months of receiving his degree, Sommerfeld asked him to write a review article on Einstein's general theory of relativity. Pauli completed the work and it was published in the German Encyclopedia of Mathematical Sciences, *where it became the authoritative reference on the subject for decades.*

Three years later he proposed that there would be a fourth quantum number, spin, and this led to the principle of exclusion for electrons. He borrowed Werner Heisenberg's matrix mathematics to show how electron spin would be represented by a 2x2 matrix, which solved the problem of how to incorporate spin into Schrödinger's mathematics for quantum states.

Wolfgang Pauli.

Pauli also discovered the neutrino in 1930. His lucid dreams were studied by Carl Jung and this led to a long series of famous letters between Jung and Pauli.

The annexation of Austria in 1938 made him a German citizen, but he fled to the United States in 1940 where he worked at the Princeton University Institute for Advanced Study until returning to Zürich in 1946 after the war.

Pauli was a perfectionist and had high expectations for anyone that worked with him. In one instance his contempt for theories that could not be tested led to his legendary dismissal of one idea as being so bad that it was 'not even wrong'. He would often find himself in a laboratory environment where his presence would seemingly cause problems with experiments in progress. This came to be known as the Pauli Effect, which he found delightful. In 1958 he died of pancreatic cancer.

Virtual particles and forces from empty space

In 1948, the Dutch physicist Hendrik Casimir proposed that what was considered to be a completely empty vacuum would nonetheless exert a force on two metal plates that would act to push them together. What was happening would only be explainable in terms of the presence of virtual photons present everywhere; everywhere except between the two metal plates. The plates were immersed in the virtual sea of photons that existed in every cubic centimetre of space, and with every possible wavelength, but the plates separated by a fixed distance, D, would cause all of the virtual photons with a wavelength equal to D to be excluded from the space between the plates. This deficit of energy between the plates would yield a force per unit area on the two plates that would push them together. There were several searches for this minuscule effect, but it wasn't until 1997 when Steve Lamoreaux at the Los Alamos National Laboratory successfully detected it with a metallic surface separated by only 0.6 to 6 microns. For greater separations, the force diminishes as the fourth-power of the plate separation ($1/D^4$) and rapidly becomes unmeasurable. For smaller separations, however, the effect continues to grow in strength. Still, there are many physical systems for which forces at this scale are important to control. One of these is the manufacture of micromachines using nanotechnology techniques. The Casimir forces produce additional sources of friction that cannot be reduced because they are literally built into the physical vacuum within which these devices operate.

The bottom line from all these discoveries is that we do, indeed, seem to live submerged in a cosmic ocean of virtual particles that cannot be directly observed. Nevertheless, the quantum world reverberates with their comings and goings in a way that can be indirectly observed. In the next chapter, we will see how this virtual world is the key to understanding one of the most curious features of the physical world: the forces that control the movement of matter.

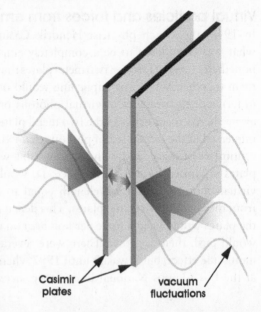

Virtual photons with long wavelengths are excluded from between the Casimir plates causing an inward pressure.

Casimir plates

vacuum fluctuations

SEARCHING FOR THE VIRTUAL VACUUM

Laboratory searches for this effect use some of the most powerful and intense lasers that can be developed today. A typical intense red laser pointer offers about 0.01 watts and produces a beam about 0.1 cm in diameter at a distance of 1 m so its intensity is about 0.01/0.01 cm^2 = 1 watt/cm^2. Calculations of the vacuum birefringence effect needs laser intensities of 10^{24} watts/cm^2 or higher. This would require, for example, a 100 petawatt (10^{17} watts) laser pulse focused on an area about 10^7 cm^2, which is an area three microns across and about 10 times smaller than a human hair. Alternatively, a relatively weak laser can also test for the effect in exceptionally strong magnetic fields. Amazingly, nature may have beaten us to this version of the experiment already.

In 2016, Roberto Mignani from INAF Milan (Italy) and a team of astronomers used European Southern Observatory's (ESO) Very Large Telescope at the Paranal Observatory in Chile to observe the neutron star RX J1856.5-3754, about 400 light years from earth. Neutron stars can have magnetic fields more than a million times stronger than any that can be created in human-scale laboratories. For this particular neutron star, its field has been measured to be at least 12 trillion Gauss. What the astronomers observed was that 16 per cent of the emitted light from the neutron star's surface was linearly polarized. This high degree of polarization would not be possible from this environment unless it was being boosted by the vacuum birefringence effect caused by the intense magnetic field. In fact, their calculations suggested that nearly 100 per cent of this measured polarization came from the vacuum effect. Future studies of this neutron star at different wavelengths will help to confirm this effect because it should get larger as the wavelength decreases, in a specific way only the quantum vacuum can produce. ESA plans to launch the Athena X-ray telescope in 2028, which should be able to verify the quantum vacuum nature of this effect once and for all. But, meanwhile, there is yet another experimental way to verify that these weird virtual particles exist: it's called the quantum Casimir effect.

Birefringence in a calcite crystal. A laser beam travels two different paths depending on its polarization.

CHAPTER 12:

Quantum Fields

The idea of a 'field' in nature is central to understanding the quantum world. Mathematically, a field is simply a quantity that varies from one location in space to another in a way that can be quantified. But the reason why, for example, gravitational fields exist and are generated by matter remained a mystery since the time Sir Isaac Newton complained about 'action-at-a-distance' in 1693. With the advent of virtual particles there was at last an explanation for why fields exist that involved the interaction of specific kinds of particles, called bosons. Particles are engulfed in clouds of fleeting virtual bosons hidden from direct view by Heisenberg's Uncertainty Principle but capable of causing observable particles to interact in specific and measurable ways. To mathematically describe how these particulate 'quantum fields' actually worked required a detailed mathematical explanation for how virtual particles interacted with matter. The initial calculations led to infinite answers even in the computation of the outcomes of very simple interactions. The discovery of the Lamb Shift in the energy of electrons within hydrogen atoms showed theoreticians that quantum field theory was not just some trivial mathematical embellishment to Dirac's relativistic quantum mechanics, but was a vital ingredient to understanding atomic physics.

Humans discover the forces of nature

For most of human history, the only force we knew about was gravity. Aristotle called it the 'tendency' that causes things made from earth (rocks, etc) to fall down towards the earth. Arabic investigators such as Abu'l-Barakāt al-Baghdādī by the 12th century described gravity in terms of acceleration. In the 17th century, Galileo showed that all forms of matter dropped from the Leaning Tower of Pisa accelerated at the same rate under gravity. Galileo's discovery allowed Sir Isaac Newton to formulate his Law of Universal Gravitation in 1666, and Newton went on in the early 1700s to explain the orbits of the planets, tides and many other phenomena. And that is where our knowledge of the gravitational force remained until Einstein's revolutionary work in 1915.

Meanwhile, the ancient Chinese knew about the curious attractive properties of magnetized lodestone, which they used as a compass. Magnetic forces were studied in detail by William Gilbert in 1600, who believed that earth was itself a giant magnet. He even showed how magnetism could be created in an iron bar by heating it and striking it with a hammer. But the detailed study of magnetism as a force did not occur until 1819 when Hans Oersted noticed that a compass accidentally placed near a wire carrying a current would twitch each time he turned the current on. By 1820, Jean-Baptiste Biot and Félix Savart, developed a mathematical formula for calculating the magnetic field around a current-carrying wire, and Michael Faraday in 1831 discovered that a changing magnetic field in a wire could induce a current to flow in another wire close by.

In an almost parallel development, the properties of charged bodies had been known since the time of the ancient Greeks. Even ancient Egyptian texts from 2750 BC make reference to electric eels as the 'Protectors of the Nile' and their ability to shock humans. Once again, William Gilbert in 1600 systematically described the difference between magnetic lodestone effects and electrostatic effects from rubbed amber. He even used the term *electron* (meaning 'amber') to describe the attractive properties of rubbed amber on some particles. Many experiments were conducted on electricity during the 17th and 18th centuries including Benjamin Franklin's iconic discovery that lightning is a form of electricity. The mathematical work on electrical phenomena began with Charles-Augustin de Coulomb proposing a law of electrostatic forces that resembled Newton's inverse-square law for gravity except that the quantity of charge on the two bodies took the place of their masses in the gravitational law. Once electrostatic forces could be studied by a mathematical law, other quantitative

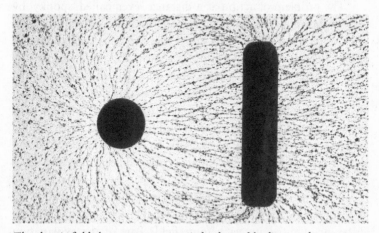

The electric fields between two oppositely-charged bodies are shown by placing the two charged objects in a pan filled with cooking oil and pepper flakes, which align in the electric field.

studies ensued in the early 1800s in a parallel development with studies of magnetism.

The unification of physics

By the mid-1800s there was one theory for magnetism and one theory for electrostatics. In the hands of James Clerk Maxwell, both of these theories were unified into one theory of electromagnetism in 1861 that provided a deep mathematical understanding of each of these phenomena. Maxwell also proposed that electromagnetic waves travelling at the speed of light were actually light waves. The final experimental connection between electromagnetism and light occurred in 1886–89 with Heinrich Hertz and his study of radio waves. So even as late as 1900, we have theories of gravity and theories of electromagnetism, but the actual natures of these forces and how matter creates them was a complete mystery. They seemed to partake of 'action-at-a-distance' but at least for electromagnetism these forces seemed rooted in the existence of an electromagnetic field: an invisible aspect to charged matter analogous to the gravitational field.

The problem of action-at-a-distance, even called 'spooky' by Albert Einstein, was a significant one for physicists all the way back to Newton's time. How was it that one body could influence the behaviour of another body at a great distance without any obvious physical contact, even when all of the air was evacuated from the intervening space? For Maxwell, this was accomplished by the interactions between the electromagnetic fields of each body, which were apparently invisible but nevertheless measurable by sensitive instruments. For gravity, Newton had no clue what a gravitational field was, only that it behaved according to a strict mathematical law.

But what exactly were these fields in-and-of themselves? Were they some kind of imperceptible fluid that leaks out of matter with different qualities for each force field? Part of the explanation was uncovered when Max Planck and Albert Einstein confirmed that electromagnetic waves (light) were a quantized phenomenon

James Clerk Maxwell.

with the light quanta later called photons by 1926. If light as an electromagnetic field had a particulate structure, the implication was that the electromagnetic field itself must be granular in a similar way. Charged particles flooded the world with clouds of photons and perhaps a similar process explained gravity as well. In 1934, the Soviet physicist Dmitrii Blokhintsev is said to have coined the term 'graviton' as the carrier particle for gravity in a research paper, 'Neutrino hypothesis and conservation of energy', but it came into much wider usage when Paul Dirac was quoted in a *New York Times* interview on 31 January 1959 as having proposed the term gravitons to describe the elementary units of gravitational waves, which were the analogues to electromagnetic waves. By the late 1920s, it seemed we were very close to a new understanding of what forces were in terms of mediating particles like photons and gravitons, but there were still problems in using this new model to mathematically calculate anything that could actually be observed.

The quantum field picture

The final clue to how force fields work as a quantum phenomenon came as the result of a conceptual marriage between Heisenberg's Uncertainty Principle (HUP), Dirac's sea of virtual particles and electron-positron pairs. The basic idea was that every particle is surrounded by a cloud of photons, but these photons are a part of Dirac's negative-energy virtual sea, which means they cannot be directly observed. It is the exchange of these 'virtual photons' that gives rise to electromagnetic forces. The reason this happens comes about because of HUP.

As we learned in Chapter 7, HUP states that there is a maximum precision with which we can ever simultaneously know measurable properties such as position, momentum and energy. The energy relationship is particularly interesting:

$$\Delta E\, \Delta t \geq \frac{\hbar}{2}$$

If we combine this with Einstein's famous equation $\Delta E = \Delta mc^2$, with a little algebra we get a new HUP relationship involving mass and time uncertainties:

$$\Delta t \geq \frac{\hbar}{2\Delta mc^2}$$

How do we interpret this? It is saying that, if we measure the mass of a system, our uncertainty Δm in being able to specify the exact value of the mass, $M \pm \Delta m$, is going to depend on our uncertainty in knowing over what time period Δt the mass was measured. In other words, the longer we continue to measure the mass of the system, the bigger Δt will be and so the smaller Δm will be. If you have two hydrogen atoms in a box, you would think that you can make a measurement of their combined mass of 1.82×10^{-30} kg with 100 per cent accuracy in a few minutes, but because hydrogen atoms are matter with wave-like properties, there is a non-zero probability that one of these atoms, or even both, might elect to leave the box by quantum tunnelling. This situation is made even worse for elementary particles such as electrons and photons.

If you take a container and remove all of the matter and energy from the box, the HUP says that only by observing this 'empty' box for billions and billions of years can you be assured that your statement 'the box is empty' is completely accurate. With Dirac's virtual sea of photons, electrons and electron-positron pairs in an unobservable, negative-energy state, HUP says that we cannot even be assured that their energies remain negative. They might actually 'pop' into our world as real positive-energy particles from time to time, but their existence as these new particles will be regulated by HUP to only a brief instant of time set by Δt. Also, the combined charges and masses for these particles have to be exactly zero because the vacuum is a state with no net charge or mass. Let's look at an example for how this works.

A photon has no net mass, but it carries an equivalent energy of E = hν thanks to Planck's Law. Suppose we consider a photon with a wavelength of 300 nm (in the ultraviolet part of the visible spectrum), which has a frequency of 1×10^{15} Hz. From Planck's Law with the constant h = 6.6×10^{-34} joules-seconds, the energy carried by this photon is just E = 6.6×10^{-19} joules. How long would we have to measure the energy of all the particles in the box before we could discern that this photon had suddenly appeared? HUP says that:

$$\Delta t \geq \frac{(6.6 \times 10^{-34})}{4\pi(6.6 \times 10^{-19})}$$

so that Δt ≥ 8.0×10^{-17} seconds. What this is saying is that a virtual photon with a wavelength of 300 nm could appear inside our box carrying an energy of 6.6×10^{-19} joules, but so long as it lived for a time *shorter* than 8.0×10^{-17} seconds, the energy it carries would not interfere with our measurement of the total energy in the box. Let's put this in a different way.

Suppose you had a very sensitive scale that you were using to measure the mass of a quantity of beach sand with a thousand sand grains in the container. It takes the instrument a full minute to settle down and give you a reading that is accurate to the mass of a single sand grain. Now suppose, unknown to you, someone dropped one additional sand grain into the box, then retrieved it before your instrument could settle down and make the precise total measurement. The coming and going of the additional sand grain would not have been detected by you, yet when it struck the other sand grains in the pile, it dislodged one of them and caused it to tumble down the pile. You would see the tumbling of these sand grains, but not be able to see the one that started the avalanche, and your mass-measuring device did not record anything more than the existing thousand sand grains in the pile.

The way that this virtual photon can affect other real particles in a system, but not be directly detected due to HUP, gives us a new idea for what fields-of-force might be. They are nothing more than clouds of virtual photons (or perhaps gravitons too), that can influence other real particles through their comings and goings, but are hidden under the cloak of the HUP, which prevents them from being directly detected. We can take this one idea a step further. Using our example of the virtual, ultraviolet photon, it can exist as a real particle for no more than 8.0×10^{-17} seconds. As a photon, it travels at the speed of light, so in this time it can travel 2.4×10^{-8} metres or 24 nm from wherever it was produced near, say, an electron. After that time, and at this distance, it has to disappear so that it does not violate the HUP Δt time constraint. If there happens to be another electron or charged particle there, it can be re-absorbed by that particle, which will then receive a 'kick' and experience the host electron's granular electromagnetic force. There are other virtual photons that can be produced too. Longer-wavelength infrared or radio photons carry much less energy and so their lifetimes will be considerably longer than this ultraviolet photon. That means their effects can travel much greater distances: centimetres, metres etc. HUP allows us to think of electromagnetic forces as simply the exchange of these virtual photons that carry a spectrum of energies and therefore have a wide range of distances over which they can influence other charged matter. This is also the reason for the Casimir Effect, which I described in the previous chapter. Another feature of this quantum model for fields is the same electron-positron pairs I mentioned earlier with the vacuum polarization effect.

According to the energy-time HUP relationship, the electron-positron pair have a total mass equal to 1.82×10^{-30} kg. This corresponds to an energy of 1.6×10^{-13} joules. From HUP this corresponds to a $\Delta t \geq 3.3 \times 10^{-22}$ seconds. So these electron-positron pairs can come and go so long as they do not linger for more than the calculated Δt, otherwise they would be detectable and violate HUP. Although they are not directly detectable, like the vacuum polarization effect, they have other measurable impacts. These should include adding slight changes to the energy of an electron inside atoms. Through the vacuum polarization effect, they should also reduce the electric charges on charged particles.

What I have described is the basic elements of forces described as quantum fields. For electromagnetic forces, they are produced by carriers called virtual photons. Presumably the virtual gravitons when exchanged produce what we experience as the force of gravity. Physicists in the 1930s found this model for forces rather compelling, but the problem they soon encountered was that there didn't seem to be an easy way to translate this picture into a mathematically-precise theory from which experimental predictions could be made.

Renormalization: what you see is not what you get

One of quantum field theory's greatest failings throughout the 1930s was its inability to give the right answer to a very simple question: what is the energy of an electron inside an atom? Each time we have improved our atomic models from Bohr to Dirac, we have added new features that have changed how we calculate the electron's energy in each of its quantum states. The last change incorporated Dirac's relativistic theory and predicted the 'hyperfine' energy splitting effect due to the electron's spin. Dirac's atomic theory is embedded in the physical vacuum and the electric fields of the protons and electrons, which are now seen as undergoing fluctuations due to the coming and going of virtual photons and electron-positron pairs. One of the first things that physicists tried to do is to calculate how minute changes in the charges of the particles caused by the vacuum polarization effect would alter the electron energy levels. These early calculations in the 1930s by many different physicists produced an embarrassing answer: these vacuum and polarization effects led to infinite energy effects, not minuscule ones. What was happening in the calculations can be seen in two different ways.

First, the electron is a point particle with no limit to how small it could be. Coulomb's Law for the electrostatic force said that the force increases as the separation decreases by D^2. If you test the strength of the electric field at half the initial distance, it becomes four times as strong. In the second view in terms of the virtual photons, this also means that their HUP travel distance is half as great, so their energy ΔE has to be twice as great. If you continue this progression down to the point-like size of the electron, the separation distance, D, vanishes, which means the Coulomb Force is now infinite, and the virtual photons are each carrying an infinite energy. Both of these conditions are completely unphysical, and so the calculations blow up and become meaningless. This 'infinity' problem raised its head every time physicists tried to calculate some feature of how electrons and photons interacted. It became the main reason why this entire approach to describing forces as quantum fields fell out of favour among physicists through much of the 1930s and 40s. It was only after World War II that circumstances for this theoretical approach changed dramatically.

While some physicists tried to create new ways of avoiding the Plague of Infinities in the new fledgling theory of quantum electrodynamics, American physicists Willis Lamb and Robert Retherford at the Columbia University Radiation Laboratory were using declassified radar technology from World War II to make high-precision measurements of the microwave spectrum of hydrogen. Because the hydrogen fine-structure line had already been investigated for confirming Dirac's new relativistic theory of the hydrogen atom, Lamb and Retherford were trying to make higher-precision measurements of the same quantum states. What they found, however, was that in addition to the hyperfine difference due to the electron spin, there was an even subtler energy difference between the $2S_{1/2}$ and $2P_{1/2}$ states by an amount 4.4×10^{-6} eV. The wavelength of the photon emitted from

Willis Lamb.

this transition is just 28.4 cm (1057 MHz) in the microwave part of the electromagnetic spectrum.

The announcement of this new energy shift was published by Lamb and Retherford on 1 August 1947. Hans Bethe at Cornell University was the first theoretician to identify this splitting as the effect of the electron emitting and absorbing virtual photons in the electromagnetic field of the proton. Unlike the earlier calculations, which gave an infinite energy splitting, Bethe took the clue from the Lamb-Retherford Shift that this quantum effect was in fact real and had a finite value. Prior to this discovery, there were no experiments that required a better model for the atom than what Dirac had provided. The Lamb-Retherford Shift provided the much-needed evidence that Dirac's theory was still incomplete and there was yet another missing ingredient. With this knowledge in hand, and with renewed motivation, Bethe developed a new way to carry out the virtual photon calculations. Bethe's paper 'The Electromagnetic Shift of Energy Levels' came out on 15 August 1947, only two weeks after the Lamb-Retherford announcement. In this paper he introduced a new calculation strategy reminiscent of what Planck had done in proposing the idea of the photon to 'cure' the infinity problem with predicting the black body spectrum at short wavelengths. Bethe decided that the concept of mass had to be altered in a process he called renormalization.

Quantum field theory

For generations, it was always assumed that the symbol, m_e, in an equation referred to the observed mass of the electron (9.1×10^{-31} kg). What physicist Hans Kramers had proposed in a lecture at the Shelter Island Conference, 2–4 June 1947, was that the observed mass m_o was actually the difference of two other masses that could be called the bare mass, m_b, and the theoretical mass m_e so that $m_o = m_e - m_b$. The bare mass of the electron was thought of as its mass without its electric field turned on. The values for m_e and m_b could each be infinite but there could also remain a very small difference, which would correspond to the actual measured mass of the electron. One way to see how this could work is by looking at mathematical series.

The two mathematical series

$$A = \sum 1/(2n-1) = 1 + \tfrac{1}{3} + \tfrac{1}{5} + \tfrac{1}{7} + \dots,$$

$$B = \sum 1/2n = \tfrac{1}{2} + \tfrac{1}{4} + \tfrac{1}{6} + \tfrac{1}{8} + \dots,$$

both diverge to infinity as you add one term to the next in their individual sums, but if you subtract A-B to get $1 - \tfrac{1}{2} + \tfrac{1}{3} - \tfrac{1}{4} + \dots$ they converge to $\ln(2) = 0.693$etc which is a finite, though irrational, number. When you think about the electron mass, m_e, and its

Linus Pauling, J. R. Oppenheimer and Richard Feynman at the Shelter Island Conference.

bare mass, m_b, these cannot be used individually in the equations without getting infinity as an answer, but if you subtract them you get the observed electron mass that appears in your experiments, m_o.

FREEMAN DYSON

Freeman John Dyson (1923–2020) was born in Berkshire, England, to a knighted composer father and a mother who was a lawyer and social worker. His older sister Alice remembered him as a boy aged four trying to calculate the number of atoms in the sun, and surrounded by encyclopedias as a voracious reader.

Between 1936–41 he attended Winchester College where his father was the director of music. At 15 he won a scholarship to Trinity College, Cambridge, where he studied mathematics, and at 19 he joined the Royal Air Force Bomber Command to work on optimizing bomber formations to maximize their effectiveness.

He returned to Trinity College and got his degree in mathematics, and in 1947 moved to Cornell University to finish his doctorate under Hans Bethe, and also crossed paths with Richard Feynman. He returned to the University of Birmingham in 1949 where he was a research fellow but by 1951 he returned to Cornell as a physics professor.

Amazingly, Dyson never received his doctorate degree for his work in physics, and although he received over 60 honorary doctorates in his life, he always regarded them as an anathema to the advancement of knowledge especially if you were a woman. Tragically, he died in 2020 from a fall at the age of 96.

Bethe, who was also attending the Shelter Island Conference, and apparently on a train ride back to Cornell University, used

Kramers' mass renormalization technique and was able to get a finite answer for the Lamb-Retherford Shift of 1040 MHz that closely matched the observation of 1057 MHz. Within a few more years, Julian Schwinger, Richard Feynman, Shin'Ichirō Tomonaga and Freeman Dyson succeeded in creating a comprehensive theory called quantum electrodynamics (QED) in which the activity among virtual particles and processes could be handled with mathematical precision. The predictions from QED are now among the most accurate of any known mathematical theory of the physical world, in some cases matching experimental results to an astonishing twelve decimal places.

Virtual photons can be converted into real photons by providing them with enough energy, such as when charged particles collide at high speeds, but are electron-positron pairs 'real'? Have they ever been observed as real particles? This has become known as the issue of 'boiling the vacuum' by experimenters or a 'QED cascade' by theorists. When you achieve electric fields as strong as 1.3×10^{18} volts/metre, called the Schwinger Limit, the quantum fluctuations in this field due to the virtual photons are strong enough that the virtual photons can themselves produce electron-positron pairs, and these pairs become real particles. Experimentally, there are two ways to do this. If you create elements with atomic numbers greater than 170, the nuclear electric fields caused by the assembled protons are strong enough that they can 'spark' the vacuum. So far, the heaviest nucleus that has been created was in 2002 called Oganesson with 118 protons, which only survived for 0.0009 seconds. A second method is to use high-power lasers, which is a technique that is evolving very rapidly. It is estimated that to cause electron-positron pairs to rain out of the vacuum, you need to collide two laser beams with opposite polarization, and with energies higher than 100 petawatts, focused to an intensity of 10^{29} watts/cm^2. At the current rate of laser power growth by 1,000x per decade, these multi-billion-dollar petawatt laser systems are becoming increasingly more expensive. Nevertheless,

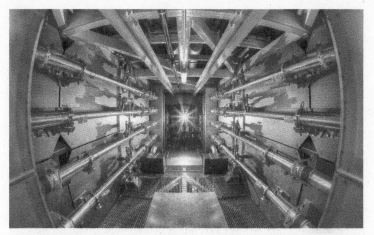

The preamplifiers of the Lawrence Livermore National Laboratory's National Ignition Facility are the first step in increasing the energy of laser beams as they make their way towards the target chamber.

at the Shanghai Super-intense Ultrafast Laser Facility in China, physicists have achieved 5 petawatts, and similar facilities at the Lawrence Livermore National Laboratory (Nova), Japan (Laser for Fast Ignition Experiments), France (PETAL) and Russia (Exawatt Center for Extreme Light Studies) are building or upgrading instruments up to 180 petawatts. It is entirely possible that the first wholesale vacuum boiling experiments may only be a decade or two away.

FEYNMAN DIAGRAMS

The detailed mathematical calculations were made simpler with the diagrammatic techniques created by Richard Feynman. With his 'Feynman diagrams' each factor in a calculation could be related to a specific line or vertex in the diagram. In fact, the diagram could be drawn first and the complex mathematical equation could be formulated just by reading out the elements of the diagram.

Feynman diagrams should be viewed with caution because they do not represent exactly what the particles are actually doing. Although electrons are shown as one-dimensional lines, they are actually waves spread out in space in accordance with HUP. Some popular accounts do not make this point, which if not noted leaves non-physicists with a misconception of how the quantum world works and how to 'see' and think about its inner structures. By analogy, the electronic schematic diagram of a radio shows symbols for resistors, capacitors and transistors, which only vaguely have anything to do with how currents flow in the circuit and how these devices actually work and interact with the electron currents.

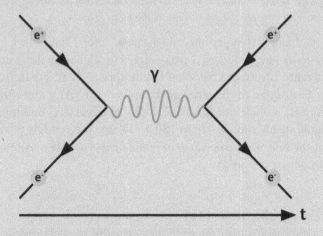

All electromagnetic interactions take place in a sea of virtual processes, which can be represented by a specific Feynman diagram.

CHAPTER 13:

The Standard Model and Beyond

For thousands of years, scientists have catalogued the basic ingredients to the physical world and the small number of elementary forces that control their motions. During the last few centuries, physicists have begun to search for underlying causes for this diversity by exploring various schemes to unify them. The first great success was James Clerk Maxwell in the mid 1800s, who combined electric and magnetic forces into a single mathematical description of electromagnetism. The growing investigations of atomic physics in the 20th century led to additional unification schemes involving quarks and leptons as simply 'fermions' and the similarities between the various 'bosons' that mediate the strong, weak and electromagnetic forces using the principles of quantum field theory. The result is now called the Standard Model, which provides the mathematical framework for making detailed predictions for every possible interaction involving these three elementary forces to exceptionally high precision compared to laboratory measurements. However, although a unified framework exists for describing the electromagnetic and weak forces, no such verified framework also includes the gluon-mediated strong interaction, which would be called a Grand Unification Theory. Including gravity leads to what is hoped to be the so-called Theory of Everything and which includes such ideas as Superstring Theory and Loop Quantum Gravity, both of which challenge our very notions of the nature of space and time themselves.

The search for unity

Just as the pyramids of Egypt had capstones made of gold-plated limestone, physicists climbing a pyramid of ideas and experiments have fully expected that the further they explore, the simpler nature will become, leading to some ultimate mathematical theory. At last one should arrive at some kind of capstone that would represent a theory so complete that it accounted for all of the details of our physical world. This has not been an idle quest because, time and again, nature has rewarded us with experimental proof that we seem to be on the right track. Using our mountain climbing analogy, our guides are leading us up a path that has now encountered some thickets that have obscured the way forward to the summit because of last year's growth. The guides have fanned out to more efficiently uncover a simpler trail upwards and we have no choice but to follow. In our ongoing story, no sooner had the physics community settled upon quantum electrodynamics as the premier theory for how electrons and photons interact via the quantum electromagnetic field, but another collection of data begged for an explanation. They had reached one important summit, but there was an even higher one just beyond. This all took place within the context of atomic physics and many investigations into the nature of radioactivity.

Discovering the third force of nature

By 1933, the electron and proton had already been discovered experimentally, but experiments by James Chadwick revealed that the atomic nucleus contained a new particle as massive as the proton (later called the neutron) but with no electric charge, leading to his receiving the Nobel Prize in Physics in 1935. Meanwhile, research into radioactivity showed that, although some elements (called isotopes) decayed to more stable elements by emitting alpha particles (helium nuclei), others decayed by emitting what were called beta rays, later identified as electrons. The 'beta decay' caused the atomic nucleus to lose one of its neutrons, which then became a proton and changed the location of the element in the periodic table. Wolfgang Pauli had studied this process and realized that there had to be missing energy carried away in the transformation, so he proposed a new elementary particle to conserve energy. The Italian-American physicist Enrico Fermi then proposed in 1933 that this invisible particle be called the neutrino, and also came up with a new interaction scheme requiring four particles to interact at a common 'vertex': a neutron, an electron, a proton and a neutrino. Now there were three fundamental forces in the universe: gravity, electromagnetism and the weak 'interaction'. But the counting of elementary forces in nature was not quite finished.

Even before the neutron was discovered, there was a severe problem with nuclear physics. The nuclear protons were all positively charged, so their mutual repulsion should have exploded every atomic nucleus in the universe. Nevertheless, atoms were stable and the universe consisted of much more than just hydrogen. What was binding the protons together in a region only 10^{-15} m in diameter? At this distance, the force or agency had to be at least 100 times stronger than the like-charge repulsion provided by the electromagnetic force. The force was also peculiar in that, like gravity, it was only attractive and faded to zero effect at distances of about 3×10^{-15} m from the nucleus. It didn't even discriminate between the charged protons and the neutral neutrons, which the force also bound tightly to nuclear dimensions. Heisenberg came up with the idea that the neutron spends part of its nuclear time as a proton-electron. The proton would be attracted to the electron causing the nuclear particles to be bound together by what he called an 'exchange interaction', which Enrico Fermi later developed into the theory of the weak interaction.

Meanwhile, in 1934, the Japanese physicist Hideki Yukawa devised a theory involving Heisenberg's exchange interaction but, not only did he calculate what the force law should look like, but in a paper published in 1935 he showed how it could be produced by an entirely new particle. At first he called it the *mesotron*, but then changed it to the more-proper Greek name *meson*, meaning 'middle one'. The meson had a mass in between that of the proton and the electron, and it was easy to calculate that it was using Heisenberg's Uncertainty Principle. As a virtual particle, HUP would show that it could only travel a distance of $\Delta x = h/4\varpi \ (mc)$, so its mass could be no more that $\Delta m = h/4\varpi c\Delta x$. With a maximum range of $\Delta x = 10^{-15}$ metres and Planck's constant 6.6×10^{-34} joules-seconds, you get $\Delta m = 2 \times 10^{-28}$ kg, which is about 190 times the mass of an electron and about $\frac{1}{10}$ the mass of a proton. The corresponding particle called the ϖ meson, or simply the pion, was finally discovered after the end of World War II in 1947 by physicists Cecil Powell, César Lattes and Giuseppe Occhialini. Yukawa was awarded the 1949 Nobel Prize for his discovery of the role played by pions in nuclear physics.

By 1950, the simple world of atomic physics consisting of the electron, proton and neutron held together by the electromagnetic force had evolved into a new arena in which four forces and a growing list of elementary particles beyond the photon and the classical atomic particles had started to emerge. There were now four fundamental forces in nature: gravity, electromagnetism, and the strong and weak interactions. The electromagnetic force was mediated by virtual photons, the strong force by virtual pions, but no one quite knew what the weak force was supposed to be mediated by according to Fermi's model. Also, the discovery of the muon and the pion in cosmic ray studies raised the number of known elementary particles to six including the photon. This became a watershed year for the discovery of a staggering number of new elementary particles, as instruments called particle accelerators or 'atom smashers' were built, beginning with the Bevatron at the Lawrence Berkeley National Laboratory, which went into operation in 1954. One year later physicists discovered the anti-proton, and a year after that the anti-neutron. Dirac's ideas about antimatter were now confirmed; every known particle had its own anti-particle. By the 1960s, the number of elementary particles had grown to several dozen, but the properties of these new sub-atomic particles were not at all random.

The quark-gluon model and the fourth force of nature

In 1964, physicists Murray Gell-Mann and George Zweig developed the idea that all of the particles that experienced the strong force, which did not include photons, electrons or muons, were not actually elementary particles at all; they were composites. To make sense of these new discoveries during what was known as the Particle Zoo Era, Gell-Mann and independently Yuval Ne'eman came up with a classification scheme for them that revealed a curious pattern when the particles were classified by their spins 0, ½ and ³⁄₂. Gell-Mann and Zweig used this geometric pattern to propose that hadrons were actually composite particles and introduced the term 'quark' as the new elementary particles. These particles were unique because they carried fractional electric charge and also had to carry a new type of charge rather arbitrarily called 'colour'. By combining these quarks three at a time you could build up all the hadron particles, and if you paired a quark with its anti-quark you could create all of the mesons. Moreover, the quark model in 1964 made a prediction for a new particle to fill a gap in the symmetry pattern for the most massive spin-$\frac{3}{2}$ hadrons. Gell-Mann predicted this particle should have an electric charge of +1 and a mass of 1.8 times that of a proton. This particle, called the Omega-minus (Ω-), was discovered using the Brookhaven accelerator in 1964. Gell-Mann received the Nobel Prize in Physics in 1969 for his development of the quark model. The quark model had to include more than quarks, because the previous carriers of the strong force were the mesons, but now they were composite particles. It would be the strong force that bound the quarks together. The nature of these new force-carrying particles was rigorously circumscribed by the way the quarks behaved as 'coloured' particles, so the force-carrying particle that bound quarks together using the colour-charge were called 'gluons'. Physicists now had a complete theory for hadrons, mesons and the strong interaction that involved quarks bound into a variety of packages of matter by the gluon, strong interaction.

The actual detection of these key quark and gluon particles took a number of years. Quarks could not be studied as free individual particles but had to be 'seen' by shooting electrons into protons and, like Ernest Rutherford had done with gold foil and alpha particles in 1911, infer the existence of quarks from the scattering data. Many of these experiments were performed at the Stanford Linear Accelerator Center in the late 1960s and generally confirmed that protons consisted of 'partons' as Richard Feynman preferred to call them. Meanwhile, the existence of gluons was confirmed in 1979 at the Deutsches Elektronen-Synchrotron in Hamburg, Germany, by observing what were called Three-Jet Events in which gluons decayed into a third spray of particles.

The weak interaction was also being studied in great detail both experimentally and theoretically, leading to an ultimate theory developed independently by Steven Weinberg, Sheldon Glashow and Abdus Salam in the late 1960s. Their model not only explained how the weak interaction functioned through the exchange of three messenger particles, the W$^+$, W$^-$ and

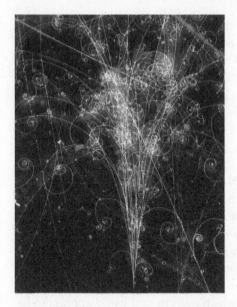

Image from the CERN bubble chamber showing a gamma ray collision (bottom) producing a spray of particles (straight lines) along with spiralling electrons and positrons.

Z^0 Intermediate Vector Bosons (IVBs). Following Yukawa's calculations, these particles would each have to be more massive than 80 protons in order for the weak interaction to have such a minuscule range of operation far smaller than an atomic nucleus. What was also remarkable was that their theory offered a way to unify the weak interaction with the electromagnetic interaction mediated by the virtual photons; called Electro-Weak Unification. This unification came about because there was an entirely new family of spin-0 particles called the Higgs bosons. When these particles interacted with the photon and the intermediate vector bosons, they caused the IVBs to gain mass but interacted so feebly with the photons that none of them gained any mass at all. In addition, these Higgs bosons interacted with all of the other quarks, gluons and leptons to confer varying degrees of mass on them as well. Gluons and photons would remain mass-free, but neutrinos would gain a smidgen of mass, electrons a bit more, and the quarks even more still. This entire unification scheme for the electromagnetic and weak interactions, in addition to the detailed mathematics for the quark-gluon system called quantum chromodynamics, led to what is now called the Standard Model beginning in the 1970s. Weinberg, Salam and Glashow shared the Nobel Prize in Physics in 1979. The Higgs boson was discovered at the Large Hadron Collider at CERN in 2012, and in 2013 Peter Higgs and François Englert received the Nobel Prize for proposing the mechanism by which particles gain mass through the Higgs interaction.

The Standard Model and its flaws

One would think that the Standard Model was the crowning achievement of 20th century physics, and in some sense it was. The inventory of the fundamental particles and forces in the universe is now complete and we have a detailed mathematical theory for how the strong, weak and electromagnetic forces operate. With the Standard Model, physicists can account for all of the phenomena upon which our physical world appears to be based, and astronomers can even account for how supernovae explode and how the Big Bang produced the universe of matter we see around us today. The mathematics, however, are fiendishly complex. They are nothing like the simple laws that Newton and Coulomb offered for gravity or electrostatic forces. They are even more complex than what James Clerk Maxwell offered for his electrodynamics.

The Standard Model seems to be complete and mathematically perfect, but there are still a number of open questions that its current version does not explain. These questions are deeply elementary and strike at the very heart of why our universe is built the way it is.

- Why are there three generations of leptons and quarks and not four or five? This limit of exactly three has even been discovered by astronomers studying the Big Bang and the cosmological formation of hydrogen and helium.

- There are 19 parameters that have to be experimentally determined, such as the masses for each particle, and the specific strengths, called the coupling constants, for the three forces. What determines these experimental values?
- What determines how many Higgs bosons exist in the Standard Model?

During the 1970s, flush with the success of electro-weak unification, physicists considered how to unify the strong force with the electro-weak; a programme of model building and theoretical work called Grand Unification Theory or simply GUT. Just as electro-weak theory could be expressed as one unified theory for these two forces when energies climbed above 300 GeV, most of the GUTs predicted that the electro-weak and the strong forces would become similar, but at the titanic energies of 1,000 trillion GeV. Moreover, there was yet another critical energy scale discovered by physicists trying to understand how to unify the three atomic and nuclear forces with gravity. This energy scale, called the Planck Scale, was 10,000 times higher than even the GUT energy scale. When physicists took a step back from their intensely mathematical work, it was eventually discovered that these theoretical efforts led to an even larger set of questions of which the most significant ones were:

- Why did nature 'decide' that our physical universe should have four energy scales: massless (gluons, photons), electro-weak (300 GeV), GUT (10^{15} GeV), and Planck (10^{19} GeV)? Why not just two energies well-defined by the maths: 'massless' and the Planck Scale?
- What determines the values of the physical constants for: the speed of light, $c = 3 \times 10^8$ m/s, the constant of gravity, $G = 6.6 \times 10^{-11}$ Ntm^2kg^{-2}, and Planck's constant, $h = 6.6 \times 10^{-34}$ j-s? These constants set the

scale for all gravitational, relativistic, and quantum phenomena in the universe. Change any one of them by 10 per cent and we would live in a very different universe.

- What is the nature of Dark Energy and Dark Matter in the universe? Astronomers have found both these phenomena in the movement of stars and galaxies. They account for 96 per cent of the gravitating 'stuff' in our universe, but so far nothing in the Standard Model accounts for them.
- Where did all the antimatter go? At every scale from cosmic to atomic, we do not live in a universe with equal amounts of matter and antimatter.

The creation of the Standard Model was a major success in simplifying a huge number of complex phenomena in the universe into a simple collection of particles, fields and equations that can literally be written on a T-shirt. But if we want to simplify things to the next level, physicists had much more work to do. As with all other research in physics, while one group was working on one set of issues to perfect the Standard Model, other groups at the same time were ploughing ahead to develop new theories beyond the Standard Model. In the Standard Model, the strong force was an entirely separate force from the electro-weak force, and gravity was still not a member of this exclusive troika. There were two grand theoretical discoveries made by these other groups that laid much of the groundwork for the last 40 years of research, which held the promise of achieving this grand synthesis, euphemistically called the Theory of Everything: supersymmetry and string theory.

The Theory of Everything

Supersymmetry, discovered by Julius Wess and Bruno Zumino in 1974, was based upon the earlier work of J.L. Gervais and B. Sakita in 1971. The Wess-Zumino application to quantum field theory was that there was a mathematical way, called a transformation, that could turn spin-½ fermions (electrons, quarks, etc) into spin-1 bosons (photons, gluons etc) but when you performed this transformation twice, the particles (fields) returned to their same initial state BUT they were shifted in space. Not only did supersymmetry simplify the Standard Model into one kind of a particle that could be either a fermion or a boson, but it seemed to include gravity as a natural part of the process. This implied that an extension of the Standard Model with supersymmetry would include unifying it with gravity, thereby serving as a prototype for unifying all of the forces in nature into one mathematical framework. There was an important catch, however. Each known particle would have to have a very massive partner to it that was at least 1,000 times as massive as a proton. These supersymmetry particles, when added to calculations involving the Standard Model, ensured that the answers would always be finite, banishing the Plague of Infinities once-and-for-all. But also, the properties of these particles were not random. They came in families like the Standard Model particles and in one of these families, called the neutralinos, the lowest-energy particle seemed to be the answer for Dark Matter.

Julius Wess, who discovered supersymmetry with Bruno Zumino in 1974.

Theoreticians during the 1990s and 2000s developed a number of extensions to the Standard Model with supersymmetry added in the simplest ways possible to create what are called the Minimally-Supersymmetric Standard Models (MSSMs). The predictions of these models have been tested at the Large Hadron Collider since 2008, but the results, so far, are not very promising. The accelerator is capable of colliding protons at energies up to 13,000 GeV. Since 2008, trillions of these collisions have been measured and studied for any signs of new, massive particles with masses above 200 GeV. Data collection experiments running for several years at a time search among the uninteresting Standard Model events for those rare occasions that would herald a new massive particle had been formed. Dozens of supercomputers at institutions around the world combine their processing power to 'solve' the identities of the particles that produce the trillions of tracks from the collision events. Although many of the tracks are merely the ordinary Standard Model particles, no sign of particles more massive than the top quark (173 GeV) have been positively detected up to the highest available energies so far. At the same time, ultra-precise measurements have been made of important physical constants related to the known Standard Model particles. The measured values are completely consistent with the Standard Model, with no 'new physics' needed such as supersymmetry. The final scientific report on the results from experiments between 2015 and 2018 delivered the dismal news; there is currently no evidence for anything other than the Standard Model and the physics it predicts. Supersymmetry is most certainly not a confirmed theory at trillion electron volt (TeV) energies where it should be making its dramatic appearance.

String theory and extra dimensions to space

Along with supersymmetry, during the late 1970s and early 1980s, physicists Michael Green and John Schwarz among many others adapted an idea from the early 1960s that particles were actually strings vibrating in many dimensions to produce the kinds of particles we observe. The first concept of a string theory for particles was created in the 1960s as an alternative to quark theory. This 'hadronic string theory' was rediscovered by Green and Schwarz but augmented by adding supersymmetry. Supersymmetry was a vital ingredient in the mathematics of string theory because it provided a means for strings that represented fermions and strings that represented bosons to be combined into one 'superstring' theory. Physicists sometimes use 'string theory' and 'superstring theory' as equivalent names because the context is understood to mean 'superstring theory'. Strings are the bedrock objects but superstring theory is the theory that describes them and includes supersymmetry. In 1984, this idea ushered in what is called the First String Revolution. Physicists soon found that if particles were strings and not infinitesimal points, many calculations would give finite answers and also gravity would play an intimate role in unifying the known particles. There were two issues that arose from this work. These strings existed in a 10-dimensional world, not the four-dimensional one we seem to live in (three dimensions of space and one dimension of time). Six of these dimensions were actually vastly smaller than atomic dimensions and not infinite. Also, these six-dimensional spaces had their own geometric features. The properties and numbers of particles in the Standard Model depended on exactly how these spaces were folded and how many 'holes' they had. These were called Calabi-Yau Manifolds because mathematicians had already studied them long before superstring theory became popular in the physics community.

Physicists were initially excited that they had at last found a candidate for the Theory of Everything. Unfortunately, a total of five different superstring theories were uncovered now that physicists knew what to look for in the mathematics. Which one uniquely represented our universe? There should, after all, only be one Theory of Everything, and that theory should specify exactly all of the observable particles, fields and the values for the physical constants we observe. In 1995 physicist Edward

Witten discovered that, if you added just one more dimension, making the world 11-dimensional, all of these five superstring theories collapsed into just one version he called M-Theory. This realization ushered in the Second String Revolution, which is the current era of theory-building. But there are no free lunches in physics. Superstring theory, although it had the capacity to explain many of the top-level questions left unanswered by the Standard Model, also had five major failings.

- It required, without any experimental evidence, that our physical world was 11-dimensional, which is far beyond the four dimensions that have been experimentally proven for millennia.
- It only made predictions for particles that had masses near the Planck energy of 10^{19} GeV.
- Each string theory offered millions of different Calabi-Yau spaces with different particles so our particular universe was lost in a landscape of possibilities and not a unique outcome.
- The theory predicted that there should be a latent energy in space similar to Dark Energy but over 10^{120} times stronger.
- It was not actually a relativistic theory because it required our four-dimensional spacetime to pre-exist in order for the strings to have a place to exist.

Superstring theory, which included the supersymmetry principle, succeeded in making a number of predictions about what the Standard Model should look like, but these were not predictions of some unknown aspect of the world. They were explanations for why the known particles and phenomena occurred as they did. By some accounts, superstring theory could be 'tuned' to give back any kind of physics you wanted. Entire universes could be conjured up by superstring theory, and our universe was not unique in any way, other than it allowed sentient life to evolve and marvel at its uniqueness from 'inside'. But the two key aspects to superstring theory that remain a problem for it are that 1) If supersymmetry is not verified experimentally, superstring theory that relies upon it is incorrect, and 2) Superstring theory does not predict what our four-dimensional 'spacetime' should look like because our spacetime is already an assumed part of the fabric on which the strings move in time and space. The first issue will eventually be settled by pursuing experiments to higher and higher energies. Eventually, supersymmetry will either be detected, or it will not. The second issue is far more problematical.

Superstring theory cannot actually be called a truly unified theory including gravity even though it predicts that graviton-like particles should exist beyond the Standard Model. The reason is that, as Einstein showed in his theory of general relativity, gravity is just another name for geometric distortions in our four-dimensional spacetime. In fact, what we call the 'gravitational field' is actually a mathematical synonym for spacetime itself. Space and time, taken together in a four-dimensional way, are components of the gravitational field itself. If you want to explain gravity, you cannot include spacetime in superstring theory from the beginning. Superstring theory has to literally create spacetime out of nothing. In fact, what all superstring theories do is to begin with a flat spacetime of four-dimensions and then add superstrings to this background. Some of these superstrings represent gravitons, but these gravitons behave as ripples of distortion in the pre-existing four-dimensional spacetime. Physicists call this a weak-field limit where the graviton particles are present but do not significantly disturb the geometry of spacetime. Because superstring theory depends on spacetime pre-existing, it cannot be a theory that obeys Einstein's principle of relativity, which is an essential feature of the Standard Model. Luckily, a new theory may have already solved this problem.

Loop quantum gravity and background independence

What is relativity? It means that there is no absolute reference frame that defines space and time as Sir Isaac Newton proposed with his Absolute Time and Space idea. The properties of space and time are only defined by making measurements with light. The result of these measurements defines a four-dimensional plenum that gives you geometric information about 3D space, and information about time. But these properties do not pre-exist your measurements. Einstein's special and general theories are relativistic because they literally 'create' our experience of time and space from measurements made between observers. No single observer is favoured and there is no single experience of time or space that is shared by all observers. Nevertheless, Einstein's theory lets us translate the experiences of one observer into the frame of reference of another observer by building up from measurements a framework that defines a consistent geometry in space and time. In 1994, Lee Smolin and Carlo Rovelli were concerned that the true meaning of a relativistic world was being lost in the current theoretical developments, so they literally went back to basics and created a theory that began with a strict relativistic foundation. It was a background-independent theory that did not presuppose that anything like four-dimensional spacetime existed. The theory would literally have to create spacetime out of even more primitive ingredients that may not look anything like space or time. This was unlike superstring theory, which had to presuppose the existence of a four-dimensional spacetime so that the strings had some place to move.

What they discovered is that there was a basic equation that had been developed in 1967 by John Archibald Wheeler and Bryce DeWitt to describe the quantum nature of gravity: the Wheeler-DeWitt Equation. It had languished in the literature for decades until Smolin realized that the solutions to this equation described curious mathematical structures called Wilson Loops.

It was an exciting moment of recognition. What this overlooked equation for quantum gravity might be describing is that our four-dimensional spacetime was in some way fashioned from these 'loops', hence this theory became known as Loop Quantum Gravity (LQG). With a bit of additional work pursuing the implications of this loopy structure for spacetime, Smolin and Rovelli discovered that underpinning these loops was a specific model for quantized spacetime. It was a collection of elementary volumes of space defined by the Planck scale of 10^{-33} cm, and these volumes were interconnected in what came to be called a spin network.

To create our spacetime, these spin networks had to be sequenced along a fourth dimension to represent how one spin network was related to the next in a series of carefully prescribed 'moves', like watching a replay of a chess game one move at a time. These four-dimensional networks were called spin foams. The exciting thing about LQG is that its natural setting is at the Planck Scale. It describes an arena one step below spacetime where elementary quantum volumes of space are interconnected, not by the equivalent of metre sticks and displacements in space, but by abstract objects that could be thought of as conduits of quantized area. The sum of these conduits or links arriving at a particular quantum volume defines the area of that volume. What do the loops have to do with this? Like tracing a pattern in a spider's web, if you encircle a specific point by tracing a closed curve around it, the number of encircled points defines the volume enclosed by this curve, and the number of links passing through the curve defines the surface area of the enclosed volume. These curves are the Wilson Loops defined by the solutions to the Wheeler-DeWitt Equation. Although the basis for background-independent LQG is far different than superstring theory, intensive work on these two theories has now identified specific questions for which LQG and superstring theory yield exactly the same answers. The most important of these common problems is the nature of the black hole singularity.

It has been known for decades that black holes are a firm prediction from Einstein's theory of spacetime called general relativity. What is predicted by this theory is that once matter is compressed inside a limiting surface called the event horizon, it is quickly crushed out of existence at the centre of the black hole in what is called a singularity. There, general relativity can no longer tell us exactly what is happening because the volume of the body has shrunk to zero and its density has climbed to infinity. Those physicists who have been working to create a quantum theory of gravity realized early on that there should exist quantum black holes vastly smaller than the diameter of an atomic nucleus, and that these also should have a singularity at their centres. In addition, the surface area of these event horizons contains information about the matter that created the black hole. In general relativity you cannot get a meaningful answer to the question of how much information can be stored in the event horizon of a quantum black hole. However, both superstring theory and LQG each provide a definite, calculable answer to this question, and the answer is precisely the same. The probability that two vastly different theories will arrive at the same answer is vanishingly small. This is viewed as evidence that perhaps BOTH theories are trying to tell us something about the Theory of Everything and how to incorporate gravity and spacetime into it.

CHAPTER 14:

Quantum Engineering

Although we can see the value in the knowledge provided by quantum mechanics for its own sake, we are, after all, tool-makers. The urge to apply any knowledge we have to making our lives more comfortable and entertaining is impossible to resist. In the 21st century, an entirely new category of careers has emerged that specifically uses the accumulated knowledge of 100 years of atomic physics: Quantum Engineering.

Engineering quantum-scale systems

Some things use quantum physics in a relatively passive way to create new technology. For instance, 'neon signs' used in commercial advertising take advantage of the quantum transitions of electrons in gases such as neon to create brilliant, colourful signage found nearly everywhere in major urban centres. There are also medical and industrial x-ray imaging systems, which take advantage of the quantum energy levels in tungsten atoms. A beam of electrons collides with tungsten atoms ejecting electrons from one of the inner quantum shells. Another electron immediately takes the place of the missing electron in the lower-energy state and emits a single photon of light at x-ray wavelengths. These so-called Roentgen Rays were used in many applications, often with deadly results.

The 1930s were a dramatic period when the growing knowledge about nuclear activity led to the discovery of the neutron and its ability to create new isotopes. Ernest Rutherford's experimentation with alpha particles from radium atoms in 1919 led him to collide them with nitrogen atoms to form oxygen: the first application of nuclear engineering. In 1932, John Cockroft bombarded atoms with artificially-accelerated protons to create new elements, and by 1935 Enrico Fermi showed that accelerated neutrons were even more effective in creating new 'radionuclides'. Atomic fission was discovered in 1938 by Otto

The first nuclear reactor.

Hahn in Germany, and that the energy produced was consistent with Einstein's iconic equation: $E = mc^2$. The first uranium fission reactor was built by Argonne National Laboratories and commissioned in 1951. By then, quantum and nuclear physics had resulted in two major industries: the production of 'atom bombs' and nuclear reactors for its peaceful uses. The 1950s were a period of world history often called the Nuclear Age because of its political and industrial transformation of our everyday life between 'duck-and-cover' drills in schools to the replacement of coal-fired plants with the 'clean' electricity from fission.

THE HIDDEN DANGERS OF SHOE SHOPPING

During the 1930s and 40s shoe stores would use x-ray imagers to help customers find the correct shoe size; an unnecessary advertising gimmick. Deaths from x-ray exposure started out as anecdotal in the 1890s, but by the 1940s it was clear that excessive exposure often led to cancers, and safety regulations were soon put into place to reduce medical technicians and industrial workers' exposure and risk.

The photoelectric effect is a bona fide quantum effect, and its utility was put to immediate use even before quantum physics was fully understood. In 1873, Willoughby Smith experimented with resistors made from the element selenium and discovered that their resistance decreased when light was applied to them, leading to Charles Fritts creating the first 'solar cell' to generate electricity directly from sunlight. Then in 1908, Karl Baedeker discovered that the conductivity of the compound copper iodide depended on the amount of iodine in the copper crystal, and defined three kinds of solid-state materials: metals, semiconductors and insulators. The race was on to understand the quantum physics of semiconductors, which had started to be used to rectify alternating current signals into direct current.

Primitive crystal set radio receivers used a cat's whisker of copper in contact with a commonly available crystal semiconductor, galena, to rectify the 'AC' signal of a radio wave into a 'DC' signal that could be used to drive a magnet in a speaker.

Almost at the same time that nuclear energy was coming into its own, the relentless evolution of electronic gadgets of every kind eventually led down the rabbit hole of the quantum world. The vacuum tubes that had so dominated the first half of the 20th century based on non-quantum physics were massive, energy-consuming devices that filled commercial television and radio receivers and many of the first generations of automatic calculating machines called computers. By the 1950s, the largest computer, the ENIAC at the University of Pennsylvania, used 20,000 vacuum tubes and required 150 kilowatts of electricity; the majority of this energy went into heating the vacuum tube filaments to create electron streams. Several of these tubes burned out every day so that the ENIAC was only operational about half the time. By the 1940s, semiconductor 'diodes' to rectify current were replacing vacuum tubes. These solid-state diodes consumed far less electricity and were much more reliable than the equivalent tubes.

The nanotechnology revolution

In 1947, William Shockley, John Bardeen and Walter Brattain used quantum physics and the growing understanding of how semiconductors work to go one step beyond the two-component 'PN Junction' diode design. They developed a 'point-contact' device using the semiconductor germanium and gold contacts on opposite sides of it. It was the first semiconductor amplifier that worked by applying a small current to the germanium to turn on or off or even modulate the flow of current between the two gold contacts. On 23 December 1947 the first sound (audio) amplifier was demonstrated at Bell Labs with a gain of about 18-times – all without using vacuum tubes. Many names were proposed for this new device including Crystal Triode and Lotatron, but the winner was the Transistor proposed by John Pierce, an American electronics engineer at Bell Labs.

Transistors slowly became commercial successes with the introduction of the hearing aid in 1953 and the transistor radio, the Regency TR-1, in 1954 at a modern-day cost of about $300. Home hobbyists could even purchase their own Raytheon CK-722 transistor for about $8. This was followed by the first commercial solid-state computer, the IBM-7070 in 1958. This computer used 30,000 transistors and could perform 27,000 operations per second at a cost of $813,000 at that time. The transistors at this time were macroscopic in size and came in metal containers 0.25-inches across. But there was nothing about the theory of transistors that required them to be this big, so manufacturers such as Texas Instruments, Motorola and Fairchild Electronics began to make them smaller and smaller. Texas Instruments took the lead in this through the invention of the 'hybrid integrated circuit' by Jack St. Clair Kilby in 1958, for which he received the Nobel Prize in 2000. Instead of large transistors soldered to a circuit board the size of a credit card, the individual components were first miniaturized and then bonded to a substrate and encapsulated into a volume the size of a sugar cube. Fully 'monolithic' integrated circuits [ICs] were developed a year later where all of the components were 'grown' on a silicon substrate by a process of lithography and deposition. Now it was just a question of how finely the lithography masks could be created to grow the various circuitry designs. The common measure of this process is the number of transistors per IC. In 1954, this was typically about 5 to 10 (transistor radio), but today it is not uncommon to find 9 billion transistors in commercial devices such as the iPhone 11 Pro smartphone. Making transistors and components more crowded than 10 billion components means that their sizes are about 7 nanometres. At this scale, we are dealing with near-atomic dimensions and this is where quantum engineering comes into its own. Even the wavelength of ordinary light is too unfocused with photolithography (500 nm) to create the exquisitely microscopic patterns, and extreme ultraviolet lithography (7 nm) or x-ray lithography (< 1 nm) has to be used instead.

The problem with designing circuitry smaller than 3 nm is that the individual wires, called 'traces', connecting the components are so small that quantum effects start to interfere with their performance as conductors. Silicon atoms are about 0.2 nm wide, so these components are only about 15 atoms wide. Quantum tunnelling effects become significant at these scales and cause electrons to leave the traces entirely; the wires become leaky. Also, transistor-based switches used in computer systems are supposed to have definite on and off states. The quantum tunnelling effects can cause them to lock up into one state. At this scale there may actually be a limit to how small we can make supercomputer systems that is set by a combination of quantum physics, the natural scale of individual atoms and Heisenberg's Uncertainty Principle. But this does not stop quantum engineers from achieving new breakthroughs at these scales.

SINGLE-ATOM SWITCHES

In 2004, quantum engineers at the Karlsruhe Institute of Technology in Germany were the first to demonstrate the action of a switch consisting of only a single atom. This technology has now been perfected and replicated by other quantum engineers around the world. The basic design is two electrodes called the source and drain between which a removable atom is placed and controlled by a small voltage applied to a 'gate' electrode. The atom is moved in and out of a gap to create an on-off switch. Because about 10,000 times less power is needed to switch the atom than in conventional nanotransistors, the development of this device may revolutionize computer and other electronics applications. Imagine carrying a modern 10-ton supercomputer in a backpack!

Conducting 'wires' only a few nanometres in diameter called 'quantum wires' were created between 1987 and 1991 and display some amazing properties. They are made from a variety of materials and consist of atoms rolled up into tubes only 10 nm in diameter or less. Electrons travel inside the wires along their central axis and experience a property known as quantum conductance, which limits the number of electrons that can travel through the wire in specific steps. The corresponding resistance step is about 26,000 ohms. Because quantum wires have very extreme aspect ratios where they can be longer than 100 microns, they are being developed for the next generation of nanotransistors, and they can also be crossed to form quantum structures called quantum dots.

Between 1978 and 1982, the Russian physicists Alexei Ekimov and Alexander Efros synthesized the first nanocrystals of copper chloride and cadmium selenide and found that these 2–10 nm crystals displayed quantum behaviour reminiscent of atoms. They had internal energy states that could be tuned by using different semiconductor compounds, which allowed the electrons in these 'quantum dots' to be stimulated to emit light at specific wavelengths. The larger quantum dots near 10 nm in size emit light mainly in the orange or red spectrum while the smaller quantum dots near 2 nm emit in the blue to green colours. These opto-electronic devices are finding their way into a number of applications such as single-electron transistors, solar cells and lasers. Their size and controllable states have also attracted attention for quantum computing and data storage. By 2008, quantum dot flash memory devices were being developed that might someday allow as much as one terabyte of data to be stored per square centimetre in 15 nm dots with 6 nanosecond retrieval times.

Manipulating atoms one at a time

Much of the excitement about nano-scale devices has involved their construction via chemical or depositional processes, but what if you could actually manipulate atoms one-by-one to place them anywhere you want and in whatever configurations? In 1981, the first scanning tunnelling microscope (STM) was developed by Gerd Binnig and Heinrich Rohrer at IBM Zurich, for which they received the 1986 Nobel Prize in Physics. They took advantage of the quantum tunnelling effect to create a conducting needle that could be scanned across the surface of any material. When the needle tip was close enough to the sample, a weak tunnelling current of electrons would flow. By arranging the instrument to always maintain exactly the same current level by automatically increasing or decreasing the height of the needle above the sample, they could map out where each atom was located in the sample at a resolution of 0.1 to 0.01 nm. Within a few years, numerous laboratories around the world were imaging the atomic surfaces of materials and even large molecules including DNA with this technique. But it soon became clear that by applying larger voltages to the tip, one could actually move atoms around on the sample surface. This was not a desirable effect for imaging so this feature of STM was actively suppressed, however the idea you could move atoms by this technique led to many nanotechnology and nanofabrication applications in the early 2000s. One of the first companies to actively investigate this technique was IBM Almaden Research Center, with their dramatic publication of their own logo at the atom-scale. This was followed in 2013 by an atom-scale movie called *A Boy and His Atom*.

Beyond the novelty of creating unusual imagery using atom manipulation, other more serious applications are also being developed. Using atom manipulation, data storage densities of one data bit per 12 atoms or about 1 terabyte per square inch can be achieved according to IBM. Another application is in manipulating electron quantum waves to also store data. By positioning atoms in a circular or elliptical ring, in 1993 IBM physicists Michael Crommie, Chris Lutz and Don Eigler demonstrated a unique technology for creating quantum

A caesium lead bromide nanocrystal under the electron microscope (crystal width: 14 nm). Individual atoms are visible as points.

interference waves at the atomic scale. These 'quantum corrals' are exquisitely interesting in terms of the basic physics alone. The scale of the corral is adjusted so it is comparable to the de Broglie wavelengths of the electrons that are present in the surface of the copper atoms in the corral's substrate. Instead of the electron waves producing a smooth interior, they follow the geometric patterns that lead to the central foci of the atomic arrangement. These locations, the central point of a circular arrangement and the two foci of an elliptical arrangement, are called quantum mirages. When a single cobalt atom was placed at one focus, the electrons near the cobalt atom arranged themselves into a specific quantum state. But amazingly, the electrons near the quantum mirage at the other focus displayed the same quantum state with no cobalt atom there. Currently, these are only useful in studying the quantum mechanics of nano-scale systems. Practical applications of quantum corrals are still rather remote because of the time-consuming fabrication and information 'readout' processes.

GERD BINNIG

Gerd Binnig was born on 20 July 1947 in Frankfurt, Germany. He received a doctorate in superconductivity from the Johann Wolfgang Goethe University in Frankfurt in 1978. He then became a research staff member at the IBM Zürich Research Laboratory. From 1985 to 1986 he worked at the IBM Almaden Research Center in San Jose, California, and from 1985 to 1988 he held a guest professorship at Stanford University. In 1987 Binnig was appointed an IBM Fellow, the highest technical position in the company.

He shared one half of the 1986 Nobel Prize in Physics with Heinrich Rohrer, which they won for their invention of the scanning tunnelling microscope (STM). Ernst Ruska won the other half of the 1986 Nobel Prize for his earlier design of the first electron microscope. Binnig and Rohrer, who also worked at the IBM laboratory in Zürich, developed the STM in the early 1980s.

Binnig's work developing the STM has earned him numerous honours in addition to the Nobel Prize, including the German Physics Prize, the Otto Klung Award and the Hewlett Packard Prize. It also stimulated an interest in creativity on his part, and his recent theoretical interests have turned towards explaining the nature of creativity and developing technologies that mimic human thought.

The development of the integrated circuit ushered in what is called the Silicon Revolution in which new applications of lithography at progressively smaller scales was pursued. The race to miniaturize a wide variety of electrical and mechanical systems especially computers and 'pocket calculators' set the stage for the modern smartphones now used by the majority of humans around the world. In physics there was also a need to come up with better and more efficient designs for accelerating particles to high energies. The current record holder is the Large Hadron Collider, a ring of thousands of superconducting magnets 27 km in circumference to accelerate particles to energies of 13,000 GeV. But this technology cannot be sustained to achieve still-higher energies.

Accelerators work by using microwave energy in hundreds of stages to increase the energies of the particles passing through specifically-designed waveguide systems several inches across. The rapidly-changing microwave electromagnetic fields boost particles to hundreds of keV per stage until they can be injected into subsequent systems to further boost their energies. Microwaves have wavelengths of several centimetres, which sets the scale for how big these systems have to be to boost particle energies, but at infrared wavelengths measured in microns, these acceleration stages can be 100,000 times smaller and still achieve similar energy boosts.

The first microdevices were the monolithic integrated circuit 'chips' developed in the late 1950s, the first micromachines to be designed, built and applied were tuning forks such as the ones developed in 1965 by Harvey Nathanson as part of an electronic device called a resonant gate transistor. While microdevices occupy the domain between about 100 and 1 microns, nanomachines are 1,000 times smaller and occupy sizes from 1 to 100 nm, with atomic size systems even smaller at between 0.01 to 1 nm. It is not uncommon to see images of micromachines such as MEMS accelerometers mis-labelled as examples of nanotechnology. True nanomachines are so small that atomic granularity can often be seen in them with STM imaging, and quantum effects such as tunnelling and the Casimir Effect have become problems.

ACCELERATORS FOR EVERYONE

In 2018, Neil Sapra and a team of researchers at Stanford University designed what was called an accelerator-on-a-chip measuring only a few centimetres across, yet capable of boosting electrons to 1 MeV energies using infrared lasers. The accelerator stages provide about 1 keV of boost over a 30 micron length (30 MeV per metre), and the stages consist of nanoscale 'mesas' separated by a channel etched out of silicon. Electrons flow through the channel and are hit by bunches of photons from an infrared laser. Although the prototype chip was only one stage, by 2025–2030 they expect to have fabricated over one thousand of these stages to give a full 1 MeV of electron energy change. Because the fabrication techniques are nearly identical to monolithic integrated circuit fabrication, they envision that millions of these inexpensive accelerators will someday be used by home hobbyists and in medicine for treating cancer and other conditions.

Atomic-scale fabrication and nanomachines

Nanotechnology and the design of devices and machines at near-atomic scale has proceeded very slowly. Because at this scale engineers are literally working with individual atoms, technologies such as STM had to appear first, followed by various proofs that the process of atom rearrangement could be controlled, hence examples of quantum corrals, IBM logos and even replicas of the Eiffel Tower.

To create simple machines, levers, motors and the equivalent of gears which are all so useful in mechanical devices, an entirely different approach has to be applied. Rather than moving individual atoms into a specific configuration, chemical assembly similar to the creation of specialized proteins has to be performed. The action of these machines has more in common with the way that DNA molecules use ribosomes to manufacture specialized proteins, which can then perform such tasks as moving chemical compounds from one location in a cell to another.

One of the first artificial mechanically-interlinked molecular structures (called an MIM) essential to fabricating molecular nanomachines was created in 1967 when Ian Thomas Harrison and Shuyen Harrison chemically synthesized a two-part molecule they called Complex-2. Dubbed a 'rotaxane' by Gottfried Schill and Hubertus Zollenkopf in 1969, this new class of MIMs consisted of an axle-like linear molecule trapped inside a second circular molecular ring. Working essentially blind because the assembly stages could only be predicted hypothetically, they arrived at the final assembly after 70 trials. Nearly 25 years later, in 1993, J. Fraser Stoddart at Edinburgh University assembled the first moving nano-wheel, called a rotaxane, whose ring could slide along the linear molecule axle from one specific station to another. This quickly led to the synthesis by Ben Feringa of an actual molecular motor capable of rotating 360° around an axle. Stoddart and Feringa went on to receive the Nobel Prize in 2016 for their pioneering work developing nanomotors.

Since the early research on fabricating nanomotors the pace of nanomachine design and fabrication has quickened. By 2017, Robert Pal at Durham University along with an international team of scientists demonstrated the first nanomotor drill that was used to destroy a prostate cancer cell. When activated by light, the motor spun three million times per second and drilled through the membrane of the cell causing instant death in a matter of a minute. The motor could bind to pre-programmed sites on the cell wall and either be used to kill the cell or to deliver a targeted compound to the cell's interior. These applications promise future applications as pharmaceuticals. Another application of nanotechnology is in energy production.

In 2010, materials scientist Zhong Lin Wang and colleagues at Georgia Institute of Technology in Atlanta used a collection of nano-wires made from zinc oxide, which is a crystal known to generate electricity when deformed. By linking 700 of these nanowires together encased in a plastic sheet, they could generate 1.26 volts. This nanogenerator technology will eventually eliminate batteries or external sources of electricity from devices as far-flung as mobile phones and micromachines. Another application of nanotechnology is in nanosensors.

In 2018, Jinming Gao and Harold Simmons at the University of Texas Simmons Cancer Center created a nanosensor that could measure the acidity of a cell and switches on to produce fluorescence that could be imaged by surgeons. Cancer cells are normally more acidotic than ordinary cells, so this nanosensor will highlight only the cancer cells for ultra-precise surgical removal. We are only at the beginning of the development of atom-sized machines, which may some day become a commonplace tool used in medicine to treat a variety of conditions.

CHAPTER 15:

Quantum Computing

While many of the nanotech discoveries and inventions mentioned in the last chapter have gone largely unnoticed by the public, one area has increasingly captured the imagination and has been promoted from many different quarters: quantum computing. While nanotechnology as a term sounds almost commonplace and mundane, placing the term 'quantum' in front of any noun automatically makes the topic sound exotic and mysterious. Today, nations are racing to become the leader in this field because of the promise of huge returns in political prestige and computational power. It has literally become the new Space Race with a world investment of about \$2 billion so far, and with China poised to invest \$10 billion in a new quantum computing centre in Hefei. So what is quantum computing? To answer this question, we first have to take a deep dive into how ordinary computers work.

Elementary computer principles

At some point in your readings you may have encountered the idea that the basis for all computers is binary mathematics consisting not of the common base-10 numbers but base-2 ones and zeros. The reason for this is simple. Computers are basically huge collections of switches that can either be on (1) or off (0). For example, the base-10 number 3 would be written as 011 and 4 would be written as 100. If you wanted to add 3+4 = 7 it would be 011+100 = 111. In each of the three slots, called 'bites', the binary adding procedure for the corresponding bites of words A and B would be accomplished by the logical operation 'A and B' where 'and' is a particular kind of electronic circuit called an AND gate. In the early history of computers, these switches and gates were vacuum tubes and mechanical relays, which were large and power-hungry. Thanks to advances in electronics and lithography, you can now crowd billions of these 'on-off' transistorized switches into a chip only a few centimetres across.

The way that computers operate is by using software to set up a series of binary operations applied to the input data to generate output data 'answers'. These answers can be as simple as a string of numbers appearing on a computer screen or printed paper, or as complex as a 3D-rendered virtual reality display. The way that most software functions is to set up a series of steps ordered in time called 'lines of coding'. This coding takes numbers from the input stream and performs operations on them in a sequential manner to arrive at an output stream, which can either be a static set of numerical answers or a movie-like simulation. As computers became more powerful in terms of the number of switches they contained, the bottom line became how fast they could manipulate numbers. The speed of a computer in operations-per-second is determined by its clock speed, which determines how fast its switches can be toggled from on to off. For humans doing pen-and-paper calculations this is about one operation per second (OPS), but the fact that you do not see the incandescent lights in your house flicker at 50 or 60 hertz means that the human visual cortex is processing information at about 50 OPS. The first commercially available Central Processing Unit (CPU) on a chip was the Intel 4004 in 1971 and it offered a speed of 92,000 OPS. It processed 4-bit data 'words' and had about 2,200 transistors with sizes of about 10 microns. As lithography improved, the density and speed of these CPUs dramatically increased throughout the 20th century.

In 1939, George Stibitz at Bell Laboratories built the first relay-based, binary adder circuit on his kitchen table to demonstrate how binary maths would work in the new generation of calculators.

The first modern desktop personal computers or 'PCs' were released in 1977: TRS-80, Apple II and the Commodore PET-2001. The TRS-80 used the Zilog-80 processor chip which used 8-bit (1 byte) data words and had a central processing unit (CPU) clock speed up to 8 million cycles per second (8 MHz), allowed for up to 8 million binary operations per second (8 MOPS). As clock speeds have steadily increased and transistors have dramatically shrunk in size, modern smartphones such as the iPhone 11 use 7 nm fabrication techniques and operate at 3 billion Hz (3 gigahertz) and through clever architecture designs can produce over 600 billion OPS. One of the chief reasons that CPUs can operate faster than their clock speeds is because of a technique discovered in the 1970s called parallel processing.

For large collections of computer code operating in a serial manner, the previous code section has to be completed before the next one can begin operation, and all of these steps run on a timeline set by the computer clock cycles. But if you take a step back and view what the entire software code is trying to accomplish, you can often break it into a series of code blocks that could be executed at any time between the start and end of the software's execution. These blocks could actually be performed by separate processors called 'cores' and their answers combined together as needed by the main program. Initially these were related to code blocks called subroutines, but later when parallel processing emerged, they were identified with separate on-chip hardware processors. There were also implementations that involved networking a number of separate computers together in what was called cluster computing. The first computers to use parallel processing architecture were built in the 1970s including the Star-100 built by Control Data Corporation for the Lawrence Livermore National Laboratory in 1974. It was the first 'pipelined, vector supercomputer', which achieved speeds of 100 million floating-point operations per second (100 megaFLOPS). The race was now on to create supercomputers of even greater speeds. The Cray 1 debuted in 1975 with a speed of 160 MFLOPS and by today the record holder is the IBM Summit built for the U.S. Department of Energy Oak Ridge National Laboratory. It can perform 148 petaFLOPS (that's 148,000,000,000,000,000 operations per second!) and uses over 2 million separate cores in a massively-parallel architecture. As for many of the larger and faster computers, they have to be cryogenically cooled because each time a switch changes its state it generates excess heat. The Summit supercomputer generates 11 billionths of a watt per operation, which at peak speed means that its collection of 2 million cores generates a staggering 800 watts of heat. Overall it requires 13 megawatts of electricity to operate, which is as much as the energy consumed by a small town of about 13,000 typical American homes.

As computers have become faster, so has the scale of the kinds of problems they can calculate. A 1 OPS system is good enough for you to do your annual taxes but is impossibly slow for calculating the positions of the planets in the sky, which was a state-of-the-art calculation in the 1800s involving human 'computers' with slide rules and pen and paper. Ballistic calculations that predicted where a launched or dropped bomb would end up became possible in real-time when speeds started to exceed about 100,000 OPS, as did the calculations for how stars evolved in time.

Not insignificantly, the new industries of computer gaming and virtual reality begun in the 1970s have placed great pressure on developing smaller and faster computers that drive the commercial development of CPUs to greater performance. Nevertheless, one of the biggest limiting factors with an entire technology based on the binary two-state switch is the heat that has to be dissipated every time a switch changes state billions of times a second. Luckily, there may be a solution to this problem, and it involves our old friend quantum mechanics. Computer architecture is based on the two-state switch that is either 100 per cent on or 100 per cent off, but in the quantum world things are not so tidy.

THE NEED FOR SPEED: ATOM BOMBS AND WEATHER FORECASTS

Two major 'hogs' for computing speed have been in weather forecasting and in the design of atomic and hydrogen bombs. In both cases, the basic laws of physics involving gravity and Newton's laws of motion have been used to make predictions in increasing detail of tomorrow's local weather and how well a particular configuration of nuclear material will produce a specific yield of devastation. In the latter instance, one no longer needs to explode an ageing nuclear warhead to determine its current megatonage. The details can be called up by doing a supercomputer model based on its configuration and radio-isotope mixtures. For weather forecasting and climatology, massive improvements have occurred as more data is being ingested, and refined codes allow simulations at near-cubic-kilometre resolution from the ground to the troposphere and beyond.

Quantum computing: bits to qubits

Ordinary computers are based on a 'classical' design created by Alan Turing in 1936 and rely on memory and algorithms that are fixed and deterministic. A specific sequence of steps will lead to a specific outcome or answer. Whether the processing is the slow-serial or lightning-fast parallel architecture, both require at their foundation logical gates and bits that are in definite electronic states, which means that a memory system consisting of n bits can be placed in 2^n different states. Also, reading out the final answers as a series of bits does not interfere with the answers. The idea of a quantum computer was first proposed by physicists Paul Benioff and Richard Feynman and differs strikingly from classical computers. The operating basis is the idea that states are defined by wave functions so that a 'bit' can be placed in a superposition of two states just as for Schrödinger's Cat. This elementary 'bit' is now called a quantum-bit or *qubit*. These mimic the information content of an electron or a photon whose spin states can be either 'up' or 'down', but before a measurement is made a qubit state is a simultaneous superposition of these. Also, qubits can be placed in entangled states just as was observed in the EPR experiment. The combination of superposition and entanglement provides two new tools for creating logic gates and doing computations that are presumably well beyond classical computing. But what can you do with a quantum computer that you can't do with an ordinary classical computer?

With an ordinary computer, if you want to determine the quickest or most efficient path connecting two different states (locations) you have to calculate all of the paths and measure for each one, say, the number of steps taken. With a quantum computer, it is placed into a superposition in which its qubits encode all of the paths at the same time. You then have to read out the quantum state to get the probabilities for the various answers, of which there will be more than one. Another application is for problems that can be expressed as a mountain climber trying to reach sea level going downhill in the fastest way possible. The quantum problem in this energy landscape would be able to take advantage of quantum tunnelling to descend along all of the possible paths to the ground state, encoding in its qubits information about the most efficient path taken. Quantum computing would also help solve many problems in molecular engineering that involve finding the final shape of a molecule based on its atomic constituents. Finally, among many other theoretical applications, are cryptology and password-cracking. These applications are hugely important to computer security and keeping information secret. For now their integrity can be preserved on classical computers. A common cryptographic technique involved encoding data using a large number that can be factored into two prime numbers called public-key encryption. Because factoring speed varies as $Log(N)$, a fast classical supercomputer could factor a $N = 150$ digit number into its primes in about a year, while a $N = 400$ digit number might take the age of the universe. This means that business and banking transactions can remain secure as well as national security information. However, some of these popular classical encryption algorithms can be broken by quantum computers rendering massive amounts of information, whose fidelity and confidentiality we rely upon, insecure.

Needless to say, the prospects for developing practical quantum computers is becoming a major investment both among private and government laboratories as the 21st century unfolds. The major technical difficulty, however, is in maintaining multiple qubits in these superimposed and entangled quantum states, and correcting them for random noise, which is always present in any system operating above absolute zero. Also, reading out the answer requires measuring the final quantum state of all the qubits. This causes the entire computer to decohere into one state unless the readout is performed carefully and in keeping with the laws of quantum mechanics.

RICHARD FEYNMAN

Richard Phillips Feynman (1918–88) was born in Queens, New York, and grew up in Far Rockaway, attended MIT as an undergraduate majoring in engineering, and later Princeton University where he received his PhD in physics in 1942.

While in secondary school, he learned how to repair radios in a laboratory he had set up in the family home. He always spoke with a thick New York accent that he seemingly maintained even during formal lectures, which his collaborators such as Wolfgang Pauli and Hans Bethe said made him sound like a 'bum'. His precocious sense of humour was also his favourite badge of honour, leading him to many avant-garde pastimes including bongo playing.

An IQ test administered to him in High School found that '125' was not high enough to admit him to the prestigious Mensa Society, however Feynman went on to score the highest rank in the Putman Mathematics Competition Exam in the United States, and in the admissions exam to Princeton University.

Following his work on the Manhattan Project, he made major contributions to quantum electrodynamics by applying path-integral techniques which led him to the development of 'Feynman Diagrams' and simple diagrammatic rules for a variety of calculations in QED. At the California Institute of Technology, he worked on the parton model for the protons (a compliment to the quark model) and issues in superconductivity. He along with Julian Schwinger and Shin'Ichirō Tomonaga was awarded the 1965 Nobel Prize in Physics for his work on the foundations of QED.

Later, Feynman uncovered the reason for the Challenger Shuttle failure in 1986 as a member of the investigatory commission. He died of liposarcoma in 1988. His last words were 'I'd hate to die twice. It's so boring.'

Richard Feynman.

Qubits of many different designs have to be cryogenically cooled to only a few millidegrees above absolute zero to minimize noise due to atomic motion in the devices. They also have to be isolated from their environment, which would tend to collapse the superposition states among the qubits. Consequently, progress has been very slow in accumulating enough qubits together to perform actual computations. Typically, a qubit consists of a nano-scale device that sets up an electron or photon in a particular spin state or superposition. The state is then stored without collapsing its wave function in a connected component called a resonator. The lifetime of the first designed qubit states was initially only a few nanoseconds but has been steadily lengthened to over 100 microseconds. Beyond these timescales, environmental contact and quantum noise causes the qubit wave function to collapse into one of its classical states in a process called decoherence. But even with decoherence times of 100 microseconds, there is still not enough time to configure a quantum computer into a state that can solve any practical problems.

Qubit arrays are kept inside cryostats and cooled to 0.1 kelvins or below. One of the largest 53-qubit arrays has recently been built by IBM for their Q System One computer. IBM has integrated this computer into its collection of other smaller-capacity quantum computers, which are collectively being used as part of a public Quantum Computation Center. By 2019 over 14 million experiments had been run by external investigators on this IBM platform. Google has also developed a 72-qubit quantum computer, but unlike IBM is only using it for in-house experiments. Compared to the billions of bits of information stored in your hard drive, 72-qubits sounds almost prehistoric but in fact is quite definitely not. The advantage of quantum computing is not in data storage but in the ability to perform parallel processing. Because a 30-qubit system can be put into

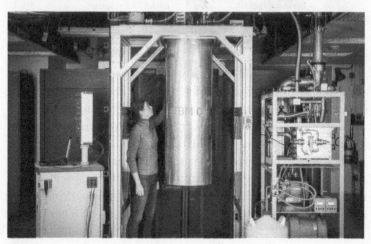

The IBM Q System One computer.

over 2^{30} different states (that's 1,073,741,824) it can perform a similar number of parallel computations at an equivalent speed of 10 teraFLOPS. A 100-qubit system could out-perform the speed of all of the supercomputers on our planet combined. A 300-qubit system could perform more calculations at once than there are atoms in the visible universe.

Quantum computing and problem solving

How fast are quantum computers? One way to assess this is to ask whether a conventional, classical computer would ever be capable of answering a question posed to a quantum computer in a reasonable amount of time. This has been termed 'quantum supremacy' – a term coined by the theoretical physicist John Preskill at CalTech. The nature of these problems is ill-defined and may not even be important in any practical way. The goal is simply to show that a classical computer cannot achieve an answer in a shorter amount of time than the same problem posed to a quantum computer. It is, however, a major technological achievement and benchmark for the entire field of quantum computing.

Some kinds of problems, however, remain beyond reach without dramatic improvements in classical computer architecture and speed. For example, real-time weather forecasting beyond the three-day horizon at high special resolution (city block or smaller) requires ingesting terabytes of ground and satellite data, computing models at minute-resolution using powerful physics-based mathematical models, and generating wind, pressure and rainfall maps in time for the next hourly radio or TV news update. Also, some other classes of research such as modelling the evolution of our universe or determining the shape of protein molecules from their chemical formulae, remain challenges near the upper limits of current supercomputer speed. A recent simulation by *Illustris* of the evolution of the universe followed 1 trillion mass points under their own gravity. The simulation took 18 million CPU hours across 8,100 cores. These are all examples of the kinds of scientific problems that require a precise knowledge of the individual steps in the computations because how a computer got to a specific answer is every bit (sorry!) as important to the scientist as the final answer itself. This is not the approach in quantum computing where the individual pathways are hidden in a mixture of superposition and entanglement and only the final answer (a cracked password or encryption code) is the computational product of interest.

COMPUTERS AT PLAY

One way in which classical computers behave in a near-quantum way is to calculate all of the possible endgames that result from a new chess move and evaluate which move to make. On 10 February 1996, IBM's Deep Blue supercomputer won its first game against a reigning chess grandmaster Garry Kasparov. Its specially-designed, massively-parallel processing capability had a speed of 11 gigaFLOPS and could evaluate 100 million moves each second. It would typically search six to 20 moves ahead to find the optimum current move. In 2011, IBM created a new question-answering computer called Watson, which competed against two previous winners on the American TV show Jeopardy. *By processing over 500 gigabytes of data per second at 80 gigaFLOPS it could answer questions considerably faster than its human competitors, winning against them after 30 minutes of play. These are examples of what one might call brute-force computing methods, which is the strength of classical computers in exhaustively (for humans!) tracking down or computing every output possibility for a given input. Modern supercomputers with their fast processors and parallel architecture are impressively fast, and for some classes of problems they excel in finding answers in a reasonable amount of time.*

On 20 September 2019, the *Financial Times* reported that 'Google claims to have reached quantum supremacy with an array of 54 q[u]bits out of which 53 were functional, which were used to perform a series of operations in 200 seconds that would take a supercomputer about 10,000 years to complete.' This claim was almost immediately contested by Google's competitors at IBM and elsewhere a few weeks later. IBM said that the same task could be performed on a classical system in just 2.5 days, rather than the 10,000 years that Google claimed. When comparing the speeds of two classical computers, it is easy to feed them the same problem, such as modelling the positions of 1 billion galaxies operating under their own self-gravity starting from an initial configuration in 3D space and velocity, but it is an entirely different matter to create difficult problems to compare a classical and quantum computer. Finding these problems that cannot be performed quickly by a classical computer, but which can also be coded for a quantum computer, are a significant contemporary challenge. The problem set up by Google was to generate and check a large number of random numbers generated by the quantum computer and verify they were random. The number of checks scales exponentially with the number of random numbers so eventually you reach a point where a classical computer slows down enormously as you increase the number of random numbers to be checked. A quantum computer does not suffer this problem in the same way and so will start to surpass the speed of a classical computer, eventually reaching the situation of quantum supremacy attained by Google. The problem identified by IBM was that the Google estimate of the time for a classical computer to perform this task did not completely account for the vast memory storage of a classical computer.

Be that as it may, we are definitely on the cusp of a new technology revolution that will have impacts and applications that we scarcely even imagine today.

CHAPTER 16:

The Nature of Matter

Having described the quantum theory of matter in great detail we have now reached something of an impasse. We have reduced the physical world into a Standard Model consisting of a small number of elementary particles and their forces, but we struggle with understanding exactly what these particles are. The issues of wave-particle duality and the violation of local Realism paint a confusing picture of exactly what particles are. The mind struggles with comprehending how localized particles can simultaneously behave as distributed wave-like objects and how it is that these objects do not even have a definite reality until they are observed. No similar objects exist in our macroworld to give us any kind of guidance as to how to form a mental picture of what an 'electron' looks like. In this chapter we review the many features of electrons that would have to be incorporated into such a model for what they are in actuality.

Particle or wave?

Winston Churchill once remarked about Russia that it was '...a riddle wrapped in a mystery inside an enigma.' No less a description is true of the electron, or any other elementary particle for that matter, cloaked as it is in its electromagnetic field, stuffed inside the equally mysterious atom. The essential dilemma is how to reconcile the idea of a localized particle with a distributed wave. In mathematics, the technique of Fourier analysis may provide a guide for understanding the particle-wave descriptions. Rather than excluding each other, they are simply different words and descriptions for the same core phenomenon.

Fourier analysis is the method of approximating one function by adding up a large number of other functions, specifically trigonometric functions such as the sine or cosine. To see how this works, the figure below shows how one kind of wave with a square shape shown in black can be decomposed into a sum of periodic waves at several different frequencies. The behaviour of the square wave in space can be translated into a set of frequencies in time for each of the component sine waves.

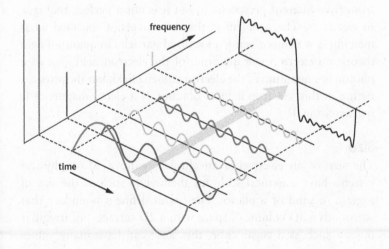

> '*The electron is for us the type of an isolated piece of energy, the one we believe, perhaps wrongly, to know best. According to received conceptions, the energy of the electron is scattered throughout the space with a very strong condensation in a region of very small dimensions whose properties are very little known to us.*'
>
> Louis de Broglie

This Fourier decomposition shows that one object (a square wave in space) described in terms of its special wavelength can be represented by another object (the component sine waves) called its frequency 'spectrum' in time. In other words, the object in the figure can be thought of as simultaneously a square wave, or a frequency spectrum. So what does this have to do with electrons?

Borrowing from the Fourier analogy, if an electron is localized in space in one description (particle with a specific radius), it can be represented in terms of a sine wave at a specific frequency or wavelength in another description (a de Broglie wave). The analogy helps us understand how we can describe a single object from two different perspectives, but it is not a perfect analogue to electrons. The problem is that we have not specified what medium is waving or being a localized particle. In quantum field theory, an electron is a quantum of the electron field, just as a photon is a quantum of the electromagnetic field, but the physical nature of this 'electron field' is obscure. Is it electromagnetic? Is it gravitational?

Size

The size of an electron is an elusive quantity. Most physical systems have a natural scale for themselves such as the size of a grain of sand or a planet. Their sizes define a boundary that surrounds a 3D volume of space with a 2D surface. For irregular objects such as a sugar cube, this scale can have many values depending on the direction from the centre. For a perfectly spherical object, only its radius is needed to establish its size in space. This does not seem to be the case for electrons. So how big is an electron?

One initial estimate, called the classical radius of the electron, can be calculated by setting the electrical potential energy of an electron equal to its relativistic rest mass energy:

$$\frac{e^2}{r} = mc^2$$

The result for m = 9.11×10^{-31} grams, e = 4.8×10^{-10} esu and c = 3.0×10^{-10} cm/s, is 2.8×10^{-13} cm. Compared to an atom with a size of 5.3×10^{-9} cm this is very small, but compared to the size of an atomic nucleus of 10^{-14} cm, this is rather large. Nevertheless, for most calculations involving the electrons in an atom, the scale of these calculations is hundreds of times larger than the electron's classical radius.

Another measure of the size of the electron is called its Compton wavelength. This is a direct application of the de Broglie equation in which the particle's Planck energy is compared to its rest mass energy.

$$\frac{hc}{\lambda} = mc^2$$

For Planck's constant h = 6.6×10^{-34} erg-sec, c = 3×10^{10} cm/s and m = 9.1×10^{-28} gm you obtain 2.4×10^{-10} cm. This is still much smaller than a typical atom-scale but significantly larger than the nuclear scale.

Experimentally, the size of an electron is generally determined by scattering other particles (photons or electrons) off of target electrons and measuring how the scattered particles behave. If the object is solid, some of these scattered particles will return along the original trajectory, but if the object is less solid, the incident particles will be deflected away from the incident direction by varying amounts depending on how deeply they penetrated the target. This was the principle used by Ernest Rutherford to discover that atoms have small, dense nuclei. From the properties of matter waves, the wavelength of an object is related to its momentum. If you scatter particles off an electron, the results will only be sensitive to features in the electron about the size of the de Broglie wavelength of the incident particle. This means that it is advantageous to use the highest energy particles you can produce for the scattering particles so they prove information at the smallest possible scales.

Precise calculations of electron-electron and electron-photon scattering by using the methods of quantum electrodynamics can predict in great detail how the incident particles will scatter for a variety of incident particle energies. This is a very sensitive calculation because the theory of QED assumes that particles can be represented as mathematical points with no physical size. By performing these scattering experiments at progressively higher energies, you can test whether for a given energy and corresponding de Broglie wavelength the electron still behaves as a point particle at that scale. If it does, and your measurements match the QED calculations, you have not reached a physical discontinuity that you could ascribe to an electron's physical size. If you do see some deviation, then this energy and the de Broglie wavelength establish an electron size at which 'new physics' has to be accounted for, including the prospect that you are getting close to an actual physical size to the electron in space. Depending on what kinds of incident particles are used, this 'new physics' region can involve the appearance of W and Z bosons, which are not covered by quantum electrodynamics. The strong interaction also makes its appearance, so that the model for scattering can become very complex as the energy of the scattering particles increases.

'we are not justified in concluding that the "thing" under examination can actually be described as a particle in the usual sense of the term... The ultimate origin of the

difficulty lies in the fact (or philosophical principle) that we are compelled to use the words of common language when we wish to describe a phenomenon, not by logical or mathematical analysis, but by a picture appealing to the imagination.'

Max Born

Scattering experiments at energies of 115 GeV involving electrons and protons at the Hadron-Electron Ring Accelerator in Hamburg, Germany, conducted in 2004 provided an upper limit of about 10^{-16} cm as the largest scale that would be consistent with the scattering data to within the measurement errors. The Large Hadron Collider, unfortunately, cannot be used to set better limits because it is a proton accelerator. However, its repeated tests of aspects of the Standard Model at energies up to 13 TeV suggests that the Standard Model, based on electrons as mathematical point particles, is still an entirely valid approximation with no detectable departures in the data compared to predictions.

Another way in which a finite size for an electron appears is through its electric dipole moment. A dipole moment is the difference between the electric charge in one hemisphere of an object and the electric charge in the opposite hemisphere multiplied by the separation between the two hemispheres, which is the radius of the object. The current upper limit value is 1.1×10^{-29} e cm measured in 2018 by the Harvard-Yale ACME II Experiment, so this indirect measure indicates no likely structure in an electron larger than about 10^{-30} cm.

Although these limits are consistent with special relativity, in general relativity there is another firm limit to the size of any object possessing mass and energy and that is its Schwarzschild radius. This is a spherical region surrounding the particle known as its event horizon, within which spacetime deforms such that light cannot escape, and whose radius is given by

$$R = \frac{2GM}{c^2}$$

For an electron with $G = 6.6 \times 10^{-8}$ Nt cm^2/gm^2, $c = 3 \times 10^{10}$ and $m = 9.1 \times 10^{-28}$ gm, we have 1.3×10^{-55} cm. This is vastly smaller than the scale at which quantum gravity theory proposes that space itself becomes quantized, called the Planck scale or 1.6×10^{-33} cm.

So, direct measurements from scattering experiments indicate a size no larger than 10^{-16} cm. Indirect measurements suggest an upper limit of 10^{-30} cm, and the theoretical Planck limit is 10^{-33} cm, so, for all intents and purposes, electrons behave as though they are mathematical points in space down to the limit where space itself may become discontinuous and quantized.

It was fashionable in 1904 to think of the electron as a small sphere carrying its charge on its surface like a planet covered with an ocean. But what was holding it together against the enormous forces of electrostatic repulsion that must surely be trying to rip it apart? Henri Poincaré offered a way out of this problem by suggesting that compensating forces within the electron would hold it together and cause it to be perfectly rigid as it moved. This idea turned out to be flawed upon closer mathematical analysis using relativity theory.

Wolfgang Pauli, meanwhile, believed that the electron's structure was one of the most pressing issues in physics. He devoted many of his early years in physics to studying it. Pauli eventually concluded that the space 'inside' an electron may in some way be physically different to the space outside it. This didn't sit too well with Einstein who was reluctant to give up the idea that space was a continuous, indivisible quantity no matter where space was located.

Mass

In classical Newtonian physics, mass is a measure of the resistance of a body to being accelerated, also termed its inertia. This is straightforward and easy to measure. All you do is apply a precise force to a body and measure its acceleration. By dividing the force by the acceleration you get a measure of its inertial mass. In 1890, Arthur Schuster measured the deflection of an electron beam in a cathode ray tube using an applied magnetic field. This gave a very precise measure of the ratio of the electron charge to its mass, Q/m. In 1909 Robert Millikan very accurately measured the charge of an electron using droplets of oil suspended by an electric field against gravity. The combination of the two measurements gave a mass for the electron of 9.1×10^{-31} kg. Because the mass of the electron appears in many spectroscopic and atomic systems, there are a multitude of different ways to measure it even more precisely leading to the current value of $9.1093837015 \times 10^{-31}$ kg, with an uncertainty of $0.0000000028 \times 10^{-31}$ kg.

In special relativity, mass is a tricky quantity to describe because the apparent mass of an object depends on its speed relative to you. However, for all observers, one thing remains an invariant and that is the rest mass of an object, which is measured

Wilhelm Wien.

in your frame of reference. Because charged bodies have electric fields, the electromagnetic field of an object also contributes to its mass via the energy of the field through $E = mc^2$.

Even before the advent of special relativity, in 1881, J.J. Thomson published a paper 'On the Effects produced by the Motion of Electrified Bodies' in which he showed that electric fields possess energy and that this contributes to the measured mass of a charged body. Experiments by Walter Kaufmann and Alfred Bucherer as early as 1901 confirmed this surprising result, but went further. The increase in the mass of a charged particle as its speed increases was identical to what had been predicted *for its electromagnetic mass alone*. What this implied, according to physicists Wilhelm Wien and Max Abraham, was that the mass of an electron, which modern tables list as 9.11×10^{-31} kg, is completely defined by its electromagnetic field. There was nothing left over in the experiments, no solid clump of Newtonian matter buried at the heart of an electron, that gave it its measured mass. The quality we identify as the mass of the electron is somehow a feature of its electromagnetic field.

An additional way in which the concept of mass becomes confusing has to do with how the electron interacts with its environment and the virtual particles in 'empty' space. In QED, the measured mass of an electron is actually a sum of two terms called the 'bare' mass and the 'electromagnetic' mass. The bare mass is what you get if you turned off the electric field of the electron, making it a neutral particle. The electromagnetic mass is the mass that the electron gains when you turn on its electric field and let it interact with the virtual particles in the vacuum. Each of these has an infinite value, but their combination is what gives you the slightly positive and finite mass that you measure to be 9.1×10^{-31} kg. Currently, because bare mass and electromagnetic mass are not measurable quantities in Nature, only the combined 'renormalized' mass is physically relevant.

In the Standard Model, all elementary particles are completely massless but gain their measured mass through an interaction of their field with the ever-present Higgs field, which is embedded in the vacuum. This means that the mass we observe is not intrinsic to the 'electron quantum field' which is massless, but comes from the interaction of this massless electron field with the Higgs field. This can be written in an equation similar to me = $\sqrt{\alpha}$ m$_H$ where α is called the Higgs-electron coupling constant and is a measure of how strongly the Higgs boson interacts with the electron. Experimentally, m$_H$ = 125 GeV or about 2.2×10^{-25} kg so the interaction strength between the Higgs boson and the electron is extremely small, and would theoretically explain why the mass of the electron is so small. Ultimately, the mass of the electron is seen as a feature of the interaction energy between these two elementary fields and from Einstein's E = mc^2 interpreted as a rest mass for the electron embedded in the Higgs field.

Quantum numbers

In addition to mass, electrons carry other bits of information. They have the intrinsic spin quantum number (½h) which means they are fermions, and they carry charge (-4.8×10^{-10} esu). The combination of a spin quantum number and a charge gives them a quantity called the magnetic moment (-9.28×10^{-24} $^1/_T$). The difficulty starts in interpreting what these quantities mean, physically.

The problem is that the so-called spin quantum number can only have a value of ½h for elementary particles and cannot be made arbitrarily larger so that this property resembles some classical property of a system. That being the case, it is actually inappropriate to think of this quantum number being a measure of how the electron is rotating about its axis like a miniature top or planet. But here's where things get even murkier.

If you take charged particles and move them in a current, that current generates a magnetic field. With the electron, we have just stated that its spin quantum number has nothing to do with the classical idea of something rotating about an axis. Nevertheless, the fact that the electron has a 'spin' and a charge seems to result in a magnetic field described by a quantity called its magnetic moment.

So the spin quantum number has nothing to do with spinning, nevertheless, the magnetic moment is a measure of a spinning charge in the classical sense. So is an electron spinning or not?

Fields

The electron has an electric field due to its charge. We saw in a previous chapter that the quantum theory of fields describes them as swarms of virtual particles that mediate the particular force they represent. Electric fields are produced by the virtual photons that interact with other charged particles to produce the electromagnetic force, which can be purely electric if the charged particle is motionless relative to you, or can produce a mixture of electric and magnetic forces if it is moving, due to the effects of special relativity.

> **THE CORRESPONDENCE PRINCIPLE**
> *One basic rule in quantum mechanics is the Correspondence Principle. This principle states that, if you take a quantum number and increase its value, you will eventually have a quantity that matches one of the classical properties of a system. For example, the energy quantum number, N, is usually very small in an atom for most normal configurations (e.g. 1,2,3... 30... 50, etc) but if you made it very large, say N = 10,000, you would end up with an atom that resembles an ordinary classical system with that energy.*

The quantum electric field surrounding an electron is layered like an onion. At great distances, the wavelengths of the virtual photons can be large and so the virtual photons carry little energy. As you get closer to the electron, the wavelengths become progressively shorter and so the virtual photons closest to the 'centre' of the electron each carry enormous amounts of energy. The sum of these energies due to the electron's self-interaction is infinite, but when the equally unobservable bare mass of the electron is added, the total mass becomes the finite value we measure as 9.1×10^{-31} kg. The graininess of the electric field is hidden from direct observation and all we experience is the smoothed classical electric field at each point in space. But exactly how and why do virtual photons elect to distribute themselves in this way about the 'centre' of an electron? The actual origin of the electric field of an electron is a mystery.

One suggestion was made in 1955 by John Wheeler, one of the early architects of quantum gravity. Taking a clue from general relativity and objects predicted by the theory called wormholes, he imagined that positive and negative charges were linked together as opposite ends of wormholes in space.

Following through with his oft-cited quote in *Gravitation* that 'What else is there out of which to build a particle except geometry (spacetime) itself?', Wheeler envisioned that space is threaded by an electric field, but that when microscopic 'quantum' wormholes form, the electric field lines get trapped and are seen to enter one of the holes and exit the other. The entering field lines are converging on one wormhole opening and the opening with its event horizon is perceived as a negatively-charged point-like particle, while the exiting field lines at the other wormhole opening are seen as a positively-charged elementary particle. He called these gravitational-electromagnetic entities 'geons' and although they could come in any size, the quantum-scale geons were ubiquitous and seen as pairs of oppositely-charged particles. Recent work on quantum entanglement has also reconsidered this idea in which pairs of entangled particles behave as opposite ends of a wormhole system. However, the original geon concept as an elementary particle does not seem to work because not only does it depend on a theory of quantum gravity to explain their details, but recent investigations by University of Victoria physicists G. Philip Perry and Fred Cooperstock in 1999 suggest that they cannot be stable.

So is an electron, or any other elementary particle, a 'particle' or a 'wave'? The answer remains indeterminate even now in the 21st century. All that we can safely say from the mathematics and experimental studies is that they are some kind of excitation in the fabric of spacetime that can be localized through certain kinds of interactions (particles). But even when other experiments show them not to be localized (waves), they still display properties such as mass, charge and spin that we associate with localized particles and not waves. To paraphrase Max Born's quote at the start of this chapter, we have to use common words to describe what we observe. These words evoke a mental picture with which we are the most comfortable. There is no deeper problem in describing these 'entities' beyond semantics.

JOHN WHEELER

John Archibald Wheeler (1911–2008) was born in Jacksonville, Florida, and grew up in Youngstown, Ohio, but between 1921–22 attended a one-room schoolhouse in Benson, Vermont, where his family had temporarily relocated.

After graduating high school in Baltimore, Maryland, in 1926 he attended Johns Hopkins University where he received his Doctorate in 1933. He then became a professor of physics at Princeton University in 1938, where he worked on models of the atomic nucleus and S-matrix theory.

He was then brought in to the Manhattan Project in 1942 and was instrumental in developing the hydrogen bomb in 1950. One of his students was Richard Feynman. Beginning in the 1950s, Wheeler developed the post-general relativity ideas of geometrodynamics, and was the key developer of many concepts in classical and quantum gravity theory, including coining the term 'black hole' and 'wormhole' along with the foamy conception of spacetime at the Planck scale. One of his most dramatic ideas was the participatory universe in which observers create the physical universe by their very act of observation.

CHAPTER 17:

The Problem with Time

Over a century after the quantum revolution we are still in the midst of trying to sort out what it all means. The granite-hard world of our ancestors has completely dissolved away into a much softer and even ephemeral world where matter has lost its solidity, an invisible virtual world steers matter, and the act of observing it all changes its reality. One of the chief casualties in all of this is the very notion of time itself. The reason why such a thing as time exists has been a source of mystery in physics for centuries. Einstein's relativity does not accord time with a deep significance in the Block Universe model, nor does it provide a reason for the existence of a definite 'now' to select from among all of the possible moments in the history of a process. Modern quantum theory considers time to be a deeply perplexing concept in physics, however recent ideas about emergent phenomena may at last provide an explanation for the origin of time.

Relativity and the Block Universe

If you look at every equation and mathematical theory describing the quantum world, they display the customary variable, t, that represents time. If you want to know what a system is doing at 5:35 UT on Friday, 10 December 2021, you just 'plug in' this time in the equation and turn the crank to get a mathematically-accurate prediction. But you could just as easily have selected 17:05 UT on Saturday, 18 January 2020, or even 09:37 on Tuesday, 15 June 1647! There is nothing in either the equations or the theoretical underpinnings of them that singles out one particular instant in time over another. This means the inescapable experience of 'now' is lost in current mathematical theories. We all have the sensation that the present moment we call Now is unique in the entire span of our lives, and even in human history itself, but there is currently no quantum theory or relativity theory that explains why Now has such a compelling character to it. In relativity theory, this is called the problem of the Block Universe. In quantum physics the equivalent idea is the collapse of the wave function after the act of measurement.

Clocks as quantum sub-systems

The Block Universe (see box below) seems complete in terms of all of the possible physical systems that can exist, but this leads to a serious problem: where is the 'clock' located that is used to indicate the ordering of the variable 't' in the cosmological equations? One of the earliest ideas for converting general relativity into a theory consistent with quantum laws was developed by John Wheeler and Bryce DeWitt. Their Wheeler-DeWitt Equation described the way that spacetime geometry states changed in an arena called superspace, but their equation had one puzzling omission: time itself. This is quite unlike quantum mechanics where Schrödinger's Equation explicitly includes time through the variable, t. To understand why this happens in relativistic models we have to look more deeply into what a universe consists of.

The underlying assumption in cosmology is that there is nothing 'outside' the universe, specifically, the spacetime of relativity is a complete object for which there is nothing external to it. All of 3D space is embedded within spacetime. When we create cosmological models using general relativity such as standard 'Big Bang' theory, it is a model of the entire contents of spacetime. But this model also contains all of the possible clocks we can invent to indicate or define a location along the time-axis. Clocks are physical sub-systems within the spacetime being modelled by cosmology. That means that the model must also describe the clock being used to define the concepts of time and 'now'. This circumstance of the clock being inside the system being modelled makes reconciling general relativity and cosmology with quantum mechanics very difficult. Quantum mechanics relies on clocks as well, but these clocks are external to the quantum system being modelled by Schrödinger's Equation. Time seems to disappear as a significant feature of cosmology and general relativity, but remains a crucial feature of quantum systems. There is another system in which time has a rather plastic and even ambiguous characteristic to it.

THE BLOCK UNIVERSE

Einstein's phenomenally successful theory of relativity states that all actions take place within the arena of a four-dimensional object called spacetime. It consists of three dimensions of ordinary space together with a one-dimensional time axis that has a geometry built up from the histories of all particles in the universe, which are individually called worldlines. Your particular worldline began at the instant of your birth and ends at the moment of your death. Your constituent atoms and molecules, however, will lead a far more complex history extending far beyond

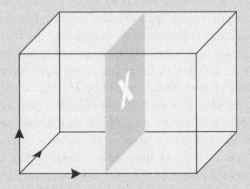

The Block Universe and its relationship to 'now' in the Einstein universe.

your birth and death. For dumb matter, these worldlines can be studied to learn about gravitational interactions and the geometric shape of the cosmos itself. An interesting feature of spacetime and worldlines is that they are fully-formed geometric objects that describe the history of individual clumps of matter all at once. Worldlines and spacetime do not evolve in time; they simply exist in a permanent and all-encompassing timeless existence. This is what physicists call the Block Universe, and in it our concept of Now is completely annihilated. Yet, if we slice this Block Universe in various ways, we recover all of the predictions from relativity about how individual particles should change from moment to moment. We recover an image of what our universe looks like today at a time 13.8 billion years after the Big Bang. So this Block Universe model has huge predictive abilities that we are willing to exchange for the fact that Now is no longer an important moment in the worldlines of local and cosmic matter.

The brain's manipulation of internal model-making and sensory data creates Now as a neuro-psychological phenomenon we experience that has a span of a few seconds, but the physical world outside our collective brain population does not operate through its own neural systems to create a Cosmic Now. That would only be the case if, for example, we were literally living inside a science fiction world resembling *The Matrix*. So in terms of physics, the idea of Now does not exist. We even know from relativity that there can be no uniform and simultaneous Now spanning large portions of space or the cosmos. If you wanted to define Now by a set of simultaneous events spanning space, relativity puts the kibosh on that idea because, due to the relative motions and accelerations of all observers, there can be no simultaneous Now that all observers can experience. Also, there is no 'flow of time' because relativity is a theory of worldlines and complete histories of particles embedded in the Block Universe of spacetime. Quantum theory, however, shows us some new possibilities.

Time as an emergent phenomenon

In previous chapters we saw that what we call 'space' may be built up like a tapestry from a vast number of events described by quantum gravity. Time may also be created from a synthesis of elementary events occurring at the quantum scale. In 1983 physicists Don Page and William Wootters discovered that the process of quantum entanglement is actually a key mechanism for creating time, which now becomes an emergent phenomenon. This is very much like what we call 'temperature' being the result of innumerable collisions among elementary objects such as atoms. Temperature is a measure of the average collision energy of a large collection of particles, but cannot be identified as such at the scale of individual particles.

A system can be described completely by its quantum state, which is a much easier thing to do when you have a dozen atoms than when you have trillions, but the principle is the same. This

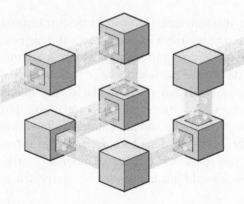

Entangled qubit networked computers are an analogue for qubits.

quantum state describes how the elements of the system are arrayed in three-dimensional space, but because of Heisenberg's Uncertainty Principle, the location of a particle at a given speed is spread out rather than localized to a definite position. A related property of quantum particles is that their states (spin, speed, position, energy, momentum) can become entangled. An intriguing set of papers by physicist Seth Lloyd at Harvard University in 1984 showed that this may be how systems evolve into an equilibrium state. Over time, the quantum states of the member particles become correlated and shared by the larger ensemble. This direction of increasing correlation goes only one way and establishes the 'Arrow of Time' on the quantum scale. The process of entanglement may also have a deep connection to the very existence of spacetime itself.

> '*There is something essential about the Now which is just outside the realm of science.*'
>
> Albert Einstein

According to Brian Swingle writing in the 2018 *Annual Review of Condensed Matter Physics*, spacetime may actually emerge through the action of quantum entanglement. Particles transcend the spacetime that separates them through the entanglement mechanism, which actually does not operate within spacetime at all. At least from theoretical considerations, researchers have found that networks of entangled states similar to the qubits used in quantum computing, create spatial correlations, but that to describe them you have to add an additional dimension to spacetime. In other words, to describe correlations in three-dimensional space, you need to add an additional fourth-dimension, which in our world just happens to be interpreted as time itself.

The implication would be equivalent to what is seen in the behaviour of holograms. A hologram encodes in its 2D surface the information needed to synthesize the image of a 3D object when exposed to a laser beam of a pure frequency. Advances in the theory of black holes, thermodynamics and information suggest that all of the information about what is inside the event horizon of a black hole is encoded in the encompassing 2D surface of its event horizon. This is similar to how quantum entanglement and the information stored in networks of qubits in 4D space are already encoded in the 3D distribution of networks of entangled qubits.

Time and human consciousness

One interesting feature of this entanglement idea is that, a few minutes ago, our brain's quantum state was less correlated with its surroundings and our sensory information than at a later time. This means that the further you go into past moments, the less correlated they are with the current brain state because, for one, the sensory information has to arrive and be processed before it can change our brain's state. Our sense of Now is the product of how past brain states are correlated with the current state. A big part of this correlation process is accomplished, not by sterile quantum interference, but by information transmitted through our neural networks that creates an internal model of our world, which is a dynamic thing.

If we did not have such an internal model that correlates our sensory information and fabricates an internal story of perception, our sense of Now would be very different because so much of the business of correlating quantum information would not occur very quickly. Instead of a Now measured in seconds, our Now would be measured in hours, and be a far more chaotic experience because it would lack a coherent, internal description of our experiences. Also, a lot can happen to an organism in an hour that does not have a survival advantage.

> *'Apparently, a geometry with the right properties built from entanglement has to obey the gravitational equations of motion. This result further justifies the claim that space-time arises from entanglement.'*
>
> Brian Swingle, 2018

Cosmologically, it is apparent from a variety of measures that events are connected in time in a definite past-present-future order. This causal ordering obtains from a variety of biological, thermodynamic and cosmological measures of closed systems changing their states generally from order to disorder following the Second Law of Thermodynamics. These are called the Arrows of Time, a term popularized by Sir Arthur Eddington in his 1928 book *The Nature of the Physical World*. We always see systems that evolve from an ordered state to a disordered one such as an ice cube melting to water, but we never see water suddenly forming a more organized ice cube.

Because it takes less information to specify an ordered state than a disordered one, physicists see a direct relationship between the entropy of a system and its information content (called its Shannon Entropy), namely, as entropy increases and the system becomes more randomized, the amount of information in the state also increases. The more improbable an event or state, the more entropy and information it contains. These issues of entropy and information content are playing a big role in the study of black holes and quantum gravity theory in which the surface area of a black hole plays the role of its entropy, and is related to the number of bits of information that can be stored on its horizon limited by the Planck scale.

Time in loop quantum gravity

Discussions about time, itself, are not found in superstring theory, which implicitly assumes the pre-existence of a special relativistic four-dimensional spacetime within which the strings exist and move. Currently, only within Loop Quantum Gravity has there been considerable attention paid to the emergence of time as a physical concept.

As discussed by Lee Smolin and Carlo Rovelli, spacetime is constructed at the quantum level from what are called spin networks. Like a vast Tinker Toy model, these networks consist of nodes that represent quantized volumes of space at the Planck scale, connected by relationships represented by edges that carry quantized units of area. At this level, space is created at the quantum level from a network of these point volumes and relationship edges. In fact, a direct analogy can be made between this network of point volumes and the network of entangled qubits previously discussed. Each of these networks represents a state of space, but these states are part of a four-dimensional network called a spin foam that represents how the linkages in one spin network rearrange themselves into another spin network along a sequence of changes. Spin foams are the quantum version of spacetime in which the fourth dimension organizing the spin networks is what we identify as time in the larger-scale 'classical' spacetime of relativity. Similarly, the holographic information stored in the three-dimensional spin networks is sufficient to build up four-dimensional spacetime.

Nevertheless, even within LQG, there is no explanation for why the organization of three-dimensional spin networks in a spin foam is treated as a time-like organization of three-dimensional things rather than merely a four-dimensional and timeless space-like structure. This is often called the 'Problem of Time' and it is an outstanding challenge for nearly all versions of quantum gravity theory. Time only appears as an internal variable that gauges how much change has occurred between the three-dimensional states of a system. This reflects the idea that there is no 'time' outside what we define as the universe, but only something that observers inside a universe experience through the use of various sub-systems called 'clocks'.

The fall of the Block Universe paradigm

This idea that no external time exists but is something created by internal configurations of matter was originally proposed in 1983 by Don Page and William Wootters, and finally put to the test by physicist Ekaterina Moreva at the Istituto Nazionale di Ricerca Metrologica in Italy. The results showed that when the polarization of an entangled system of two photons was measured no changes occurred, because the properties (polarization) of both photons was measured outside the entangled system. This is the perspective of a cosmic 'super-observer'. However, when the polarization of one photon was measured from within the system, the observer becomes entangled with it. When this measurement is compared with the polarization of the second photon also within the system, the difference is a measure of time. This confirms that time is an emergent property of an entangled system as perceived by 'clocks' within a universe, but not by an external clock.

Experiments such as these at the quantum level suggest that the Block Universe perspective of relativity is not correct, and that although the past can be reconstructed from essentially classical physics and records of stored information (photographs etc), the future is closely determined by probabilities and principles found in quantum mechanics. The 'present' is when the quantum mechanical probabilities become 'crystallized' into the certainty of the past. Physicist George Ellis in 2009 proposed this as the concept of the Crystallizing Block Universe, which is very consistent with what LQG founder Lee Smolin proposes.

The idea that time is not a fundamental property of nature but an emergent one is similar to the concept of temperature, which is well defined for large collections of particles, but is meaningless for individual particles. Other examples of emergent phenomena include the well-known properties of liquid water, turbulence, air pressure, rainbows, life and even consciousness itself. Closely related to emergence is the idea of self-organization. Here we have individual objects (birds) operating collectively by applying a simple rule (stay close to your neighbour; when they move you move too). As 'murmurations' of starlings show, this can lead to beautiful and complex patterns in time and space.

A murmuration of starlings.

LEE SMOLIN

Lee Smolin was born in 1955 in New York City and received his PhD in physics from Harvard University in 1979. Smolin held many faculty positions at Yale, Pennsylvania State and Syracuse University before becoming a Visiting Scholar at the Princeton Institute for Advanced Study in 1995. In 2001 he became one of the founding faculty members of the Perimeter Institute in Waterloo, Ontario.

Smolin had been interested in quantum gravity issues since the late 1980s and worked with Carlo Rovelli, Abhay Ashtekar and Ted Jacobson to formulate such a theory called Loop Quantum Gravity based on purely relativistic principles with no pre-existing spacetime. Smolin has also been an active popularizer, and investigator, of both experimental and philosophical consequences of quantum gravity theory. His popular accounts of the work on quantum gravity include the influential The Trouble with Physics, *which identified the major failings of string theory as a true 'theory of everything'. His recent 2019 book* Einstein's Unfinished Revolution *is also considered to be pathbreaking and thought-provoking as it explores what lies beyond a quantum description of the world.*

Another mechanism for the emergence of time is via a process called quantum tunnelling (see page 89). At the atomic and nuclear scale, spontaneous transitions appear in the decay or fissioning of certain nuclei. For example, a Polonium-212 nucleus spontaneously emits a helium nucleus in a process called alpha decay. The time you have to wait for this for a collection of polonium nuclei is about 0.2 microseconds. The escape of the alpha particle from the polonium nucleus is not permitted classically because it violates the conservation of energy. The alpha particle just doesn't have enough energy to break free from the nucleus but escape is permitted quantum mechanically because of Heisenberg's Uncertainty Principle. In other words, you can't really tell if the alpha particle at a specific location in the nucleus has exactly the energy it needs to escape. The time needed to escape the nucleus and 'tunnel' through the energy barrier depends on the energy difference. The bigger the energy difference, the longer it will take in an exponential way. A controlled application of quantum tunnelling can be found in the Scanning Tunnelling Microscope, which is able to detect individual atoms in a number of different systems by detecting the electron tunnelling current.

According to Stephen Hawking, a similar tunnelling process may have occurred at the Big Bang. In its initial state, which could have been similar to the spacelike attributes of the spin networks in LQG, cosmic spacetime may have been in a four-dimensional, pure space state. The spin foam we discussed previously may have been a purely four-dimensional space-like object. But then through a tunnelling event, one of the space-like dimensions tunnelled into a time-like dimension and, quite literally, time began in what we now call the Big Bang. From then on one might suppose that the quantum entanglement process created events and sub-systems called 'clocks' from which state changes would be interpreted as ongoing, time-like changes between states. Taking Hawking's idea one step further, the tunnelling event may have been irregular so that some regions of the pre-existing four-dimensional space may have remained unaffected while other 'bubbles' of the true spacetime may have formed with a time-direction. By whatever means this event came about, we were left with a universe deeply connected to the quantum world, and one that came into being with a definite set of physical laws, whose details set the stage for the eventual emergence of sentient life itself.

Index

Picture Credits

American Computer & Robotics Museum, Bozeman, Montana: 88b

Alamy: 13, 28t, 40, 63, 75, 79, 134, 135l, 168t

CalTech: 55l (LIGO), 55r (LIGO), 142t (CalTech Archives)

CERN: 8b, 110, 117, 146, 156

Cornell University Library: 142b

Daniel Barredo: 172b (Paris-Saclay, Institut d'Optique & CNRS)

David Woodroffe: 24, 34, 62, 70, 81, 109t, 153, 171, 173b, 199t, 199b

ESO: 131, 136

Flickr: 174 (IBM), 181b (IBM)

Getty Images: 18b, 20t, 59, 67, 72, 98, 100, 114, 141, 165t, 182

Instituto de Astrofísica de Andalucía: 118 (CSIC)

Library of Congress: 31, 127, 190t

Microbe TV: 18t

NASA: 32, 33 (x2), 52t, 54, 119b (GSFC), 120 (ESA), 139, 140 (ESA)

NOAO: 91

Science Museum: 28bl

Science Photo Library: 8t, 9, 19, 21, 35tl, 35b, 82, 86, 109, 116r, 119t, 152, 155, 160, 168b, 170, 172t, 173t, 199t, 200

Shutterstock: 12, 28br, 29b, 30l, 35tr, 36, 38, 41, 42, 43, 44t, 44b, 46, 48l, 48r, 56, 58, 61, 64, 74, 76, 89, 102 (Debby Wong), 103, 122, 124, 132, 138, 144, 148 (akimov Konstantin), 162, 184, 187, 190b, 194, 203

Wellcome Collection: 10, 17t, 22, 25, 26, 27r, 39, 51, 85, 88, 135r

Wikimedia Commons: 16, 17b, 27l, 29t, 49, 78, 90, 92b, 104, 115, 116l, 120, 125, 126, 129, 130t, 130b, 143, 145 (Damien Jemison/LLNL), 164, 165b, 166, 168, 178, 180, 181t, 183, 197, 198